Behaviormetrics: Quantitative Approaches to Human Behavior

Volume 14

Series Editor

Akinori Okada, Professor Emeritus, Rikkyo University,
Tokyo, Japan

This series covers in their entirety the elements of behaviormetrics, a term that encompasses all quantitative approaches of research to disclose and understand human behavior in the broadest sense. The term includes the concept, theory, model, algorithm, method, and application of quantitative approaches from theoretical or conceptual studies to empirical or practical application studies to comprehend human behavior. The Behaviormetrics series deals with a wide range of topics of data analysis and of developing new models, algorithms, and methods to analyze these data.

The characteristics featured in the series have four aspects. The first is the variety of the methods utilized in data analysis and a newly developed method that includes not only standard or general statistical methods or psychometric methods traditionally used in data analysis, but also includes cluster analysis, multidimensional scaling, machine learning, corresponding analysis, biplot, network analysis and graph theory, conjoint measurement, biclustering, visualization, and data and web mining. The second aspect is the variety of types of data including ranking, categorical, preference, functional, angle, contextual, nominal, multi-mode multi-way, contextual, continuous, discrete, high-dimensional, and sparse data. The third comprises the varied procedures by which the data are collected: by survey, experiment, sensor devices, and purchase records, and other means. The fourth aspect of the Behaviormetrics series is the diversity of fields from which the data are derived, including marketing and consumer behavior, sociology, psychology, education, archaeology, medicine, economics, political and policy science, cognitive science, public administration, pharmacy, engineering, urban planning, agriculture and forestry science, and brain science.

In essence, the purpose of this series is to describe the new horizons opening up in behaviormetrics — approaches to understanding and disclosing human behaviors both in the analyses of diverse data by a wide range of methods and in the development of new methods to analyze these data.

More information about this series at https://link.springer.com/bookseries/16001

Nobuoki Eshima

An Introduction to Latent Class Analysis

Methods and Applications

Springer

Nobuoki Eshima
Department of Pediatrics and Child Health
Kurume University
Kurume, Fukuoka, Japan

ISSN 2524-4027 ISSN 2524-4035 (electronic)
Behaviormetrics: Quantitative Approaches to Human Behavior
ISBN 978-981-19-0971-9 ISBN 978-981-19-0972-6 (eBook)
https://doi.org/10.1007/978-981-19-0972-6

This Springer imprint is published by the registered company Springer Nature Singapore Pte Ltd.
The registered company address is: 152 Beach Road, #21-01/04 Gateway East, Singapore 189721, Singapore

Preface

In observing human behaviors and responses to various stimuli and test items, it is valid to assume they are dominated by factors, for example, attitudes and abilities in Psychology, and in Sociology, belief, social customs and folkways, and so on; however, it is rare to observe and measure such factors directly. The factors are hypothesized components that we employ in scientific researches, and the factors are unobservable and treated as latent variables to explain phenomena under consideration. The factors are sometimes called latent or internal factors. Although the latent factors cannot be measured directly, data of the observable variables, such as responses to test items and interviews with respect to national elections, are obtainable. The variables are referred to as the manifest variables. Based on the observations, the latent factors have to be estimated, and in order to do it, latent structure analysis was proposed by Lazarsfeld (1950). For extracting latent factors, it is needed to collect multivariate data of manifest variables, and in measuring the variables, then, we have to take response errors into consideration, which are those induced from physical and mental conditions of subjects, intrusion (guessing) and omission (forgetting) errors, and so on. When observing results of a test battery from examinees or subjects, it is critical how their abilities can be assessed and our interest is how we order the examinees according to the latent factors instead of simple scores, that is, sums of item scores, by using the test battery. Latent structure analysis is classified into latent class analysis, latent trait analysis and latent profile analysis according to the types of manifest and latent variables, in a strict sense. Latent class analysis treats discrete (categorical) manifest and latent variables; latent trait analysis deals with discrete manifest variables and continuous latent variables; and latent profile analysis handles continuous manifest variables and discrete latent variables. The purpose of latent structure analysis is similar to that of factor analysis (Spearman, 1904), so in a wide sense, factor analysis is also included in latent structure analysis. Introducing latent variables in data analysis is ideal; however, it is sensible and meaningful to explain the phenomena under consideration by using latent variables. In this book, latent class analysis is taken up in the focus of discussion, and applications of latent class models to data analyses are treated in the several themes, that is, exploratory latent class analysis, confirmatory latent class analysis, analysis of longitudinal data, path analysis with latent class models and so on. Along with it, latent profile and latent trait models are also treated in the parameter estimation. The author would like to expect the present book to play a significant role in introducing latent structure analysis to not only young researchers and students studying behavioral sciences, but also those investigating in the other scientific research fields.

Kurume, Japan Nobuoki Eshima

References

Lazarsfeld, P. F. (1950). The logical and mathematical foundation of latent structure analysis. In S. A. Stuffer, L. Guttman, and others (Eds.), *Measurement of Prediction: Studies in Social Psychology in World War II*, **4**, Princeton University Press.

Spearman, S. (1904). "General-intelligence", objectively determined and measured. *American Journal of Psychology*, **15**, 201–293

Acknowledgements

I would like to express my sincere gratitude to Prof. Yushiro Yamashita, chairman of the Department of Pediatrics & Child Health, Kurume University School of Medicine, for providing me with excellent environments and encouragements to complete this book. I would also be very much indebted to Dr. Shigeru Karukaya for his useful advice to throw myself into finishing this book.

Contents

Chapter 1
Overview of Basic Latent Structure Models

1.1 Introduction

Latent structure analysis is classified into four analyses, i.e., latent class analysis, latent profile analysis, latent trait analysis, and factor analysis. Latent class analysis was introduced for explaining social phenomena by Lazarsfeld [11], and it analyzes discrete (categorical) data, assuming a population or group under study is divided into homogeneous subgroups which are called latent classes. As in the latent class model, assuming latent classes in a population, latent profile analysis (Gibson, 1959) was proposed for the study of interrelationships among continuous variables. In this sense, the model may be regarded as a latent class model [1, 8]. Latent trait analysis has been developed in a mental test theory (Lord, 1952, Lord & Novic, 1968) and also employed in social attitude measurements [10]. The latent trait model was designed to explain responses to manifest categorical variables depending on latent continuous variables, for example, ability, attitude, and so on. The systematic discussion on the above models was given in Lazarsfeld & Henry [9]. Factor analysis dates back to the works of Spearman [20], and the single factor model was extended to the multiple factor model [21]. The analysis treats manifest and latent continuous variables and explains phenomena under study by extracting simple structures to explain inter-relations between the manifest and latent variables. Although "latent structure analysis" is now a general term for the analyses with the above models, in many cases, the name is used for latent class analysis in a narrow sense after Lazarsfeld [11]. In the early years of the developments of the latent structure models, the main efforts on studies were placed on parameter estimations by solving the equations with respect to the means and covariances of manifest variables, which are called the accounting equations; however, now, the methods are only in the historical development. As the efficiency of computers has been increased rapidly, these days, the method of maximum likelihood (ML) can be easily applied to data analyses. Especially, the expectation–maximization (EM) algorithm [3] provided a great contribution to parameter estimation in latent structure analysis.

N. Eshima, *An Introduction to Latent Class Analysis*, Behaviormetrics: Quantitative Approaches to Human Behavior 14,
https://doi.org/10.1007/978-981-19-0972-6_1

In this chapter, latent structure models are reviewed. Section 1.2 treats the latent class model and the accounting equations are given. In Sect. 1.3, a latent trait model with discriminant and item difficulty parameters is discussed. A comparison of the model and a latent class model is also made. Section 1.4 treats the latent profile model, which is regarded as a factor analysis model with categorical factors, and in Sect. 1.5, the factor analysis model is briefly reviewed. Section 1.6 reviews generalized linear models (GLM) [15, 16], and latent structure models are treated in a GLM framework, and in Sect. 1.7, the EM algorithm for the ML estimation of latent structure models is summarized. Finally, in Sect. 1.8, a summary and discussions of the present chapter are provided.

1.2 Latent Class Model

In the latent class model, it is assumed a population is divided into some subpopulations, in which individuals are homogeneous in responses to items under study. The subpopulations are called latent classes in the analysis. Let X_i be manifest variables that take categories $\{1, 2, \ldots, K_i\}$, $i = 1, 2, \ldots, I$, which imply the responses to be observed, and let ξ be a latent variable that takes categories $\{1, 2, \ldots, A\}$, which denote latent classes and they are expressed by integers for simplicity of the notations. Let $\boldsymbol{X} = (X_1, X_2, \ldots, X_I)^T$ be the I-dimensional column vector of the manifest variables X_i; let $P(\boldsymbol{X} = \boldsymbol{x}|a)$ be the conditional probability of $\boldsymbol{X} = \boldsymbol{x}\left(= (x_1, x_2, \ldots, x_I)^T\right)$ for given latent variable (class) a; and let $P(X_i = x_i|a)$ be those of $X_i = x_i$. Then, in the latent class model, conditional probability $P(\boldsymbol{X} = \boldsymbol{x}|a)$ of $\boldsymbol{X} = \boldsymbol{x}$ for a given latent variable (class) a is expressed by

$$P(\boldsymbol{X} = \boldsymbol{x}|a) = \prod_{i=1}^{I} P(X_i = x_i|a). \tag{1.1}$$

The above equation indicates that manifest variables X_i are statistically independent in latent class a. The assumption is called that of local independence. Let v_a be the probability that a randomly selected individual in a population is from latent class a. Then, from (1.1), we have

$$P(\boldsymbol{X} = \boldsymbol{x}) = \sum_{\alpha=1}^{A} v_a P(\boldsymbol{X} = \boldsymbol{x}|a) = \sum_{\alpha=1}^{A} v_a \prod_{i=1}^{I} P(X_i = x_i|a), \tag{1.2}$$

where

$$\sum_{\alpha=1}^{A} v_a = 1; \sum_{x_i=1}^{K_i} P(X_i = x_i|a) = 1, i = 1, 2, \ldots, I. \tag{1.3}$$

Table 1.1 Positive response probabilities of latent class model (1.4)

Latent class	X_1	X_2	$\cdots$	X_I
1	π_{11}	π_{12}	$\cdots$	π_{1I}
2	π_{21}	π_{22}	$\cdots$	π_{2I}
$\vdots$	$\vdots$	$\vdots$	$\vdots$	$\vdots$
A	π_{A1}	π_{A2}	$\cdots$	π_{AI}

The above equations are referred to as the accounting equations. In many data analyses, manifest variables are binary, for example, binary categories are {yes, no}, {positive, negative}, {success, failure}, and so on. Such responses are formally denoted as integers {1,0}. Let π_{ai} be the positive response probabilities of manifest variables $X_i, i = 1, 2, \ldots, I$. Then, the Eqs. (1.2) are expressed as follows:

$$P(\boldsymbol{X} = \boldsymbol{x}) = \sum_{\alpha=1}^{A} v_a \prod_{i=1}^{I} \pi_{ai}^{xi} (1 - \pi_{ai})^{1-xi}. \tag{1.4}$$

According to the above accounting Eqs. (1.2) and (1.4), the latent class model is also viewed as a mixture of the independent response models (1.1). The interpretation of latent classes is done by latent response probabilities $(\pi_{a1}, \pi_{a2}, \ldots, \pi_{aI})$ (Table 1.1). Exploratory latent class analysis is performed by the general model (1.2) and (1.4), where any restrictions are not placed on model parameters v_a and $P(X_i = x_i|a)$. On the other hand, in confirmatory analysis, some constraints are placed on the model parameters. The constraints are made according to phenomena under study or the information of practical scientific research.

Remark 1.1 In latent class model (1.2) with constraints in (1.3), the number of parameters v_a is $A - 1$ and that of $P(X_i = x_i|a)$ is $A \sum_{i=1}^{I}(K_i - 1)$. Since the number of manifest probabilities (parameters) $P(\boldsymbol{X} = \boldsymbol{x})$ is $\prod_{i=1}^{I} K_i - 1$, in order to identify the latent class model, the following inequality has to hold:

$$\prod_{i=1}^{I} K_i - 1 > (A - 1) + A \sum_{i=1}^{I} (K_i - 1). \tag{1.5}$$

Remark 1.2 In the latent class model (1.2), single latent variable ξ has been assumed for explaining a general framework of the model. In a confirmatory latent class analysis, some latent variables can be set for the analysis, for example, an application of the latent class model to explain skill acquisition patterns [2] and latent class factor analysis model [14]; however, in such cases, since latent variables are categorical and finite, the models can be viewed as restricted cases of the general latent class model. For example, for a latent class model with two latent variables ξ_j with categorical sample spaces $\left\{1, 2, \ldots, A_j\right\}, j = 1, 2$, setting $(\xi_1.\xi_2) = (a, b)$ as a new latent variable $\zeta = a + A_1(b-1), a = 1, 2, \ldots, A_1, b = 1, 2, \ldots, A_2$, the model can be viewed as a restricted case of the general model.

1.3 Latent Trait Model

Let θ be the latent trait (ability) of a randomly selected individual in a population, where the latent trait is a real value or vector in a Euclidian space; let X_i be manifest variables that take categories $\{1, 2, \ldots, K_i\}, i = 1, 2, \ldots, I$ as in Sect. 1.2; let $P(\boldsymbol{X} = \boldsymbol{x}|\theta)$ be the conditional probabilities of responses $\boldsymbol{X} = \boldsymbol{x}\left(= (x_1, x_2, \ldots, x_I)^T\right)$, given latent trait θ; and let $P(X_i = x_i|\theta)$ be those of $X_i = x_i$. Under the assumption of local independence, the latent trait model is given by

$$P(\boldsymbol{X} = \boldsymbol{x}|\theta) = \prod_{i=1}^{I} P(X_i = x_i|\theta),$$

where $P(X_i = x_i|\theta), i = 1, 2, \ldots, I$ are real-valued functions of θ. Let $\varphi(\theta)$ be the standard normal density function of latent trait $\theta \in (-\infty, +\infty)$, i.e.,

$$\varphi(\theta) = \frac{1}{\sqrt{2\pi}} \exp\left(-\frac{\theta^2}{2}\right).$$

Then, we have

$$P(\boldsymbol{X} = \boldsymbol{x}) = \int_{-\infty}^{+\infty} \varphi(\theta) \prod_{i=1}^{I} P(X_i = x_i|\theta) d\theta. \tag{1.6}$$

Comparing (1.2) and (1.6), model (1.6) can be approximated by a latent class model. Let

$$-\infty\left(=\theta_{(0)}\right) < \theta_{(1)} < \theta_{(2)} < \ldots < \theta_{(A-1)} < +\infty\left(=\theta_{(A)}\right), \tag{1.7}$$

and let

$$v_a = \int_{\theta_{a-1}}^{\theta_a} \varphi(\theta) d\theta, a = 1, 2, \ldots, A. \tag{1.8}$$

Then, (1.6) can be approximated as follows:

$$P(\boldsymbol{X} = \boldsymbol{x}) \approx \sum_{\alpha=1}^{A} v_a \prod_{i=1}^{I} P(X_i = x_i | \theta_a), \tag{1.9}$$

where we set

$$\theta_{(a-1)} < \theta_a < \theta_{(a)}, a = 1, 2, \ldots, A.$$

For binary manifest variables, positive response probabilities $P_i(\theta)(= P(X_i = 1|\theta))$ are non-decreasing functions. For example, the two-parameter logistic model [12] is given by

$$P_i(\theta) = \frac{1}{1 + \exp(-Da_i(\theta - d_i))} = \frac{\exp(Da_i(\theta - d_i))}{1 + \exp(Da_i(\theta - d_i))}, i = 1, 2, \ldots, I, \tag{1.10}$$

where a_i and d_i are discriminant and difficulty parameters, respectively, for test item $i, i = 1, 2, \ldots, I$, and $D = 1.7$. This model is an extension of the (Rasch model (1960)) and popularly used in item response theory. In general, the above functions are referred to as item characteristic functions. Positive response probabilities $P_i(\theta)$ are usually continuous functions in latent trait θ (Fig. 1.1).

Remark 1.3 Let Y_i be latent traits to answer items $i, i = 1, 2, \ldots, I$ and let θ be a common latent trait to answer all the items. It is assumed variables Y_i and θ are jointly distributed according to a bivariate normal distribution with mean vector (0, 0) and variance–covariance matrix,

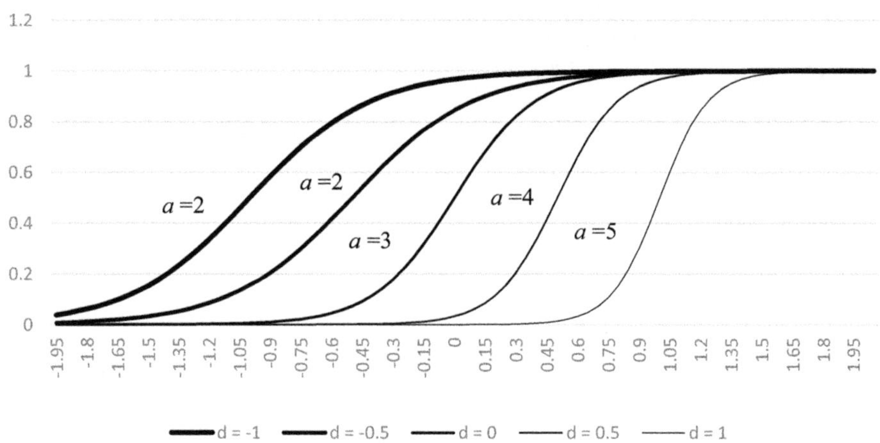

Fig. 1.1 Two-parameter logistic models (1.10)

$$\Sigma = \begin{pmatrix} 1 & \rho_i \\ \rho_i & 1 \end{pmatrix}.$$

From this, the conditional distributions of Y_i for given θ are normal $\mathrm{N}\left(\rho_i\theta, 1 - \rho_i^2\right)$. Let η_i be the threshold of latent ability Y_i to successfully answer to item $i, i = 1, 2, \ldots, I$. The probabilities that an individual with latent trait θ gives correct answers to items $i, i = 1, 2, \ldots, I$ are computed as follows:

$$P_i(\theta) = P(Y_i > \eta_i | \theta) = \Phi\left(\frac{\rho_i}{\sqrt{1 - \rho_i^2}}\left(\theta - \frac{\eta_i}{\rho_i}\right)\right), i = 1, 2, \ldots, I,$$

where $\Phi(x)$ is the standard normal distribution function. Setting

$$a_i = \frac{\rho_i}{\sqrt{1 - \rho_i^2}}, d_i = \frac{\eta_i}{\rho_i}, D = 1.7,$$

we have

$$\Phi(a_i(\theta - d_i)) \approx \frac{\exp(Da_i(\theta - d_i))}{1 + \exp(Da_i(\theta - d_i))}.$$

The treatment of the logistic models in both theoretical and practical discussion is easier than that on the normal distribution model $\Phi(a_i(\theta - d_i))$, so the logistic models are used in item response models. The graded response model [18, 19] is an extension of this model. ■

In order to improve the Guttman scale model, the latent distance model [9] was proposed by using step functions. Let θ be a latent trait on interval [0, 1] and let the thresholds $\theta_{(i)}, i = 1, 2, \ldots, I$ be given as

$$\left(\theta_{(0)} =\right)0 < \theta_{(1)} < \theta_{(2)} < \ldots < \theta_{(I)} < 1\left(= \theta_{(I+1)}\right). \tag{1.11}$$

Then, the item characteristic functions are defined by

$$P_i(\theta) = \begin{cases} \pi_i^L \ \left(\theta < \theta_{(i)}\right) \\ \pi_i^H \ \left(\theta \geq \theta_{(i)}\right) \end{cases}, i = 1, 2, \ldots, I, \tag{1.12}$$

where $0 \leq \pi_i^L < \pi_i^H \leq 1$. Probabilities π_i^L imply guessing errors and $1 - \pi_i^H$ forgetting ones. In the above model, thresholds (1.11) imply the difficulties of items as well. If we set

$$v_a = \theta_{(a)} - \theta_{(a-1)}, a = 1, 2, \ldots, I + 1, \tag{1.13}$$

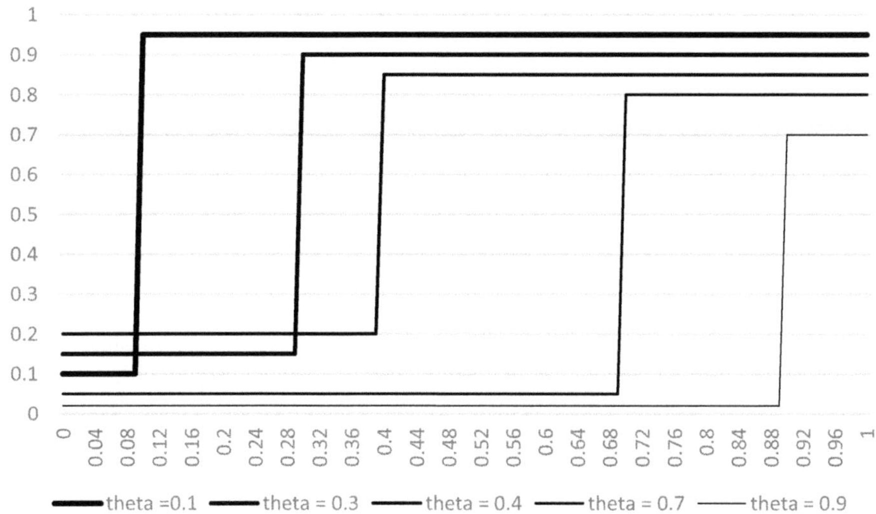

Fig. 1.2 Latent distance model (1.12)

the latent distance model is a restricted version of the latent class model. As in the two-parameter logistic model, the graphs of $P_i(\theta)$ are illustrated in Fig. 1.2.

1.4 Latent Profile Model

Assuming that a population under study is decomposed into A subpopulations, for continuous manifest variables $X_i, i = 1, 2, \ldots, I$, let $f(\boldsymbol{x}|a)$ and $f_i(x_i|a)$ be the conditional density functions of $\boldsymbol{X} = (X_1, X_2, \ldots, X_I)^T$ and X_i in latent class (subpopulation) $a(= 1, 2, \ldots, A)$, respectively; and let v_a be proportions of latent classes $a(= 1, 2, \ldots, A)$. Then, under the assumption of local independence, it follows that

$$f(\boldsymbol{x}|a) = \prod_{i=1}^{I} f_i(x_i|a). \tag{1.14}$$

Hence, the joint density function of $\boldsymbol{X}$, $f(\boldsymbol{x})$, is given by

$$f(\boldsymbol{x}) = \sum_{a=1}^{A} v_a \prod_{i=1}^{I} f_i(x_i|a), \tag{1.15}$$

where

$$\sum_{a=1}^{A} v_a = 1.$$

If the conditional density functions $f_i(x_i|a)$ are normal with mean vectors μ_{ai} and variances $\psi_i^2, i = 1, 2, \ldots, I$, then, from (1.14), we have

$$f(\boldsymbol{x}|a) = \prod_{i=1}^{I} \frac{1}{\sqrt{2\pi\psi_i^2}} \exp\left\{-\frac{(x_i - \mu_{ai})^2}{2\psi_i^2}\right\}. \tag{1.16}$$

This model can be expressed with linear equations, and in latent class a, the following equations are given:

$$X_i = \mu_{ai} + e_i, i = 1, 2, \ldots, I, \tag{1.17}$$

where e_i are the error terms that are independently distributed according to normal distributions with means 0 and variances $\psi_i^2, i = 1, 2, \ldots, I$. From the above equations, the variances and covariances of the manifest variables, $\sigma_i^2 = \mathrm{Var}(X_i)$ and $\sigma_{ij} = \mathrm{Cov}(X_i, X_j)$, are described as follows:

$$\begin{cases} \sigma_i^2 = \sum_{a=1}^{A} v_a (\mu_{ai} - \overline{\mu}_i)^2 + \psi_i^2 \ (i = 1, 2, \ldots, I), \\ \sigma_{ij} = \sum_{a=1}^{A} v_a (\mu_{ai} - \overline{\mu}_i)(\mu_{aj} - \overline{\mu}_j) \ (i \neq j), \end{cases} \tag{1.18}$$

where

$$\mathrm{E}(X_i) = \overline{\mu}_i = \sum_{a=1}^{A} v_a \mu_{ai}, i = 1, 2, \ldots, I. \tag{1.19}$$

Figure 1.3 illustrates the mixed distribution of normal distributions $N(-2, 1)$ and $N(5, 1)$, where the mixed ratios are 0.6 and 0.4, respectively. The number of parameters in model (1.16) is $I(A + 2) - 1$ and that of the above manifest equations is $\frac{1}{2}I(I + 3)$. From this, in order to identify the model, the following equation has to hold:

$$I(A + 2) - 1 < \frac{1}{2}I(I + 3). \tag{1.20}$$

The latent profile model can be viewed as a factor analysis model with categorical factors.

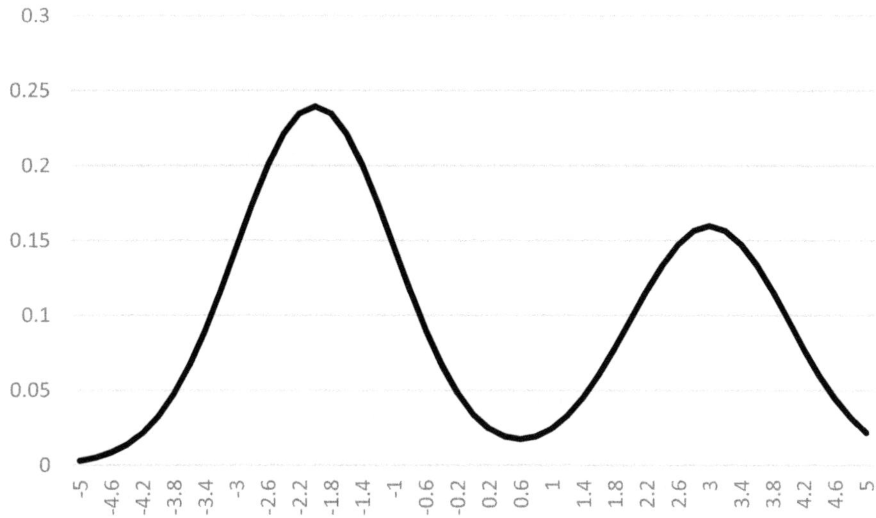

Fig. 1.3 The density function of $0.6N(-2, 1) + 0.4N(5, 1)$

1.5 Factor Analysis Model

Let X_i be manifest variables, ξ_j latent variables (common factors), and ε_i unique factors peculiar to X_i; and let λ_{ij} be factor loadings that are weights of factors ξ_j to explain X_i. Then, the factor analysis model is given as follows:

$$X_i = \sum_{j=1}^{m} \lambda_{ij}\xi_j + \varepsilon_i, i = 1, 2, \ldots, I,$$

where

$$\begin{cases} \mathrm{E}(X_i) = \mathrm{E}(\varepsilon_i) = 0, i = 1, 2, \ldots, I; \\ \mathrm{E}(\xi_j) = 0, j = 1, 2, \ldots, m; \\ \mathrm{Var}(\xi_j) = 1, j = 1, 2, \ldots, m; \\ \mathrm{Var}(\varepsilon_i) = \psi_i^2 > 0, i = 1, 2, \ldots, I; \\ \mathrm{Cov}(\varepsilon_k, \varepsilon_l) = 0, k \neq l. \end{cases}$$

Assuming that factors $\xi_j, j = 1, 2, \ldots, m$ and $\varepsilon_i, i = 1, 2, \ldots, I$ are normally distributed, then, the conditional density functions of manifest variables of $X_i, i = 1, 2, \ldots, I$ given the factors $\xi_j, j = 1, 2, \ldots, m$ are described by

$$f_i(x_i|\boldsymbol{\xi}) = \frac{1}{\sqrt{2\pi\psi_i^2}}\exp\left(-\frac{\left(x_i - \sum_{j=1}^{m}\lambda_{ij}\xi_j\right)^2}{2\psi_i^2}\right).$$

The conditional normal density function of $\boldsymbol{X}$ given $\boldsymbol{\xi}$ is expressed as

$$f(\boldsymbol{x}|\boldsymbol{\xi}) = \prod_{i=1}^{I}\frac{1}{\sqrt{2\pi\psi_i^2}}\exp\left(-\frac{\left(x_i - \sum_{j=1}^{m}\lambda_{ij}\xi_j\right)^2}{2\psi_i^2}\right). \tag{1.21}$$

Comparing (1.16) and (1.21), the latent profile model can be viewed as a factor analysis model with categorical factors.

In latent class analysis, the latent class factor analysis model is briefly reviewed [14]. Let $\xi_j, j = 1, 2, \ldots, m$ be binary latent variables; let $X_i, i = 1, 2, \ldots, I$ be binary manifest variables; let $f(\boldsymbol{x}|\boldsymbol{\xi})$ be the conditional probability function of $\boldsymbol{X} = (X_1, X_2, \ldots, X_I)^T$ given $\boldsymbol{\xi} = (\xi_1, \xi_2, \ldots, \xi_m)^T$. Assuming there are no interactions between the latent variables, then, the model is expressed as follows:

$$f(\boldsymbol{x}|\boldsymbol{\xi}) = \prod_{i=1}^{I}\frac{\exp\left(\alpha_i + x_i\sum_{j=1}^{m}\lambda_{ij}\xi_j\right)}{1 + \exp\left(\alpha_i + x_i\sum_{j=1}^{m}\lambda_{ij}\xi_j\right)}. \tag{1.22}$$

In a factor analysis model, the predictor in the above model is $\sum_{j=1}^{m}\lambda_{ij}\xi_j$ and regression coefficients λ_{ij} are log odds with respect to binary variables X_i and ξ_j, and the parameters are interpreted as the effects of latent variables ξ_j on manifest variables X_i, that is, the positive response probabilities are changes by multiplying $\exp(\lambda_{ij})$. The latent class factor analysis model (1.22) is similar to the factor analysis model (1.21).

1.6 Latent Structure Models in a Generalized Linear Model Framework

Generalized linear models (GLMs) are widely applied to regression analyses for both continuous and categorical response variables [15, 16]. As in the above discussion, let $f_i(x_i|\boldsymbol{\xi})$ be the conditional density or probability function of manifest variables X_i given latent variable vector $\boldsymbol{\xi}$. Then, in GLMs, the function is assumed to be the following exponential family of distributions:

$$f_i(x_i|\boldsymbol{\xi}) = \exp\left(\frac{x_i\theta_i - b_i(\theta_i)}{a_i(\varphi_i)} + c_i(x_i, \varphi_i)\right), i = 1, 2, \ldots, I, \tag{1.23}$$

where θ_i and φ_i are parameters and $a_i(\varphi_i)(>0)$, $b_i(\theta_i)$ and $c_i(x_i, \varphi_i)$ are specific functions for response manifest variables $X_i, i = 1, 2, \ldots, I$. This assumption is referred to as the random component. If X_i is the Bernoulli trial with $\mathrm{P}(X_i = 1) = \pi_i$, then, the conditional probability function is

$$f_i(x_i|\boldsymbol{\xi}) = \pi_i^{x_i}(1-\pi_i)^{1-x_i} = \exp\left(x_i \log\left(\frac{\pi_i}{1-\pi_i}\right) + \log(1-\pi_i)\right),$$

Corresponding to (1.23), we have

$$\theta_i = \log\left(\frac{\pi_i}{1-\pi_i}\right), a(\varphi_i) = 1, b(\theta_i) = -\log(1-\pi_i), c(x_i, \varphi_i) = 0. \tag{1.24}$$

In this formulation, for binary manifest variables, latent class model (1.1) can be expressed as follow:

$$\begin{aligned} P(\boldsymbol{X} = \boldsymbol{x}|a) &= \prod_{i=1}^{I} \exp\left(x_i \log\left(\frac{\pi_{ai}}{1-\pi_{ai}}\right) - \log(1-\pi_{ai})\right) \\ &= \exp\left(\sum_{i=1}^{I} x_i \theta_{ai} - \sum_{i=1}^{I} \log(1-\pi_{ai})\right) \\ &= \exp\left(\boldsymbol{x}^T \boldsymbol{\theta}_a - \sum_{i=1}^{I} \log(1-\pi_{ai})\right), \end{aligned} \tag{1.25}$$

where

$$\boldsymbol{\theta}_a^T = (\theta_{a1}, \theta_{a2}, \ldots, \theta_{aI}), \theta_{ai} = \log\left(\frac{\pi_{ai}}{1-\pi_{ai}}\right), i = 1, 2, \ldots, I.$$

For normal variable X_i with mean μ_i and variance φ_i^2, the conditional density function is

$$f_i(x_i|\boldsymbol{\xi}) = \frac{1}{\sqrt{2\pi\psi_i^2}} \exp\left(\frac{x_i\mu_i - \frac{1}{2}\mu_i^2}{\psi_i^2} + \left(-\frac{x_i^2}{2\psi_i^2}\right)\right),$$

where

$$\theta_i = \mu_i, a_i(\varphi_i) = \psi_i^2, b_i(\theta_i) = \frac{1}{2}\mu_i^2, c_i(x_i, \varphi_i) = -\frac{x_i^2}{2\psi_i^2} - \log\left(\sqrt{2\pi\psi_i^2}\right).$$

In the factor analysis model (1.21), the random component is reformulated as follows:

$$f(\boldsymbol{x}|\boldsymbol{\xi}) = \prod_{i=1}^{I} \exp\left(\frac{x_i\theta_i - \frac{1}{2}\theta_i^2}{\psi_i^2} + \mathrm{c}(x_i, \psi_i^2)\right) = \exp\left(\sum_{i=1}^{I} \frac{x_i\theta_i - \frac{1}{2}\theta_i^2}{\psi_i^2} + \sum_{i=1}^{I} \mathrm{c}(x_i, \psi_i^2)\right). \tag{1.26}$$

Let us set

$$\boldsymbol{\Psi} = \begin{pmatrix} \psi_1^2 & & & \\ & \psi_2^2 & 0 & \\ & 0 & \ddots & \\ & & & \psi_I^2 \end{pmatrix}, \boldsymbol{\theta}^T = (\theta_1, \theta_2, \ldots, \theta_I), \boldsymbol{x}^T = (x_1, x_2, \ldots, x_I).$$

Then, (1.26) is re-expressed as follows:

$$f(\boldsymbol{x}|\boldsymbol{\xi}) = \exp\left(\boldsymbol{x}^T \boldsymbol{\Psi}^{-1}\boldsymbol{\theta} - \frac{1}{2}\boldsymbol{\theta}^T \boldsymbol{\Psi}^{-1}\boldsymbol{\theta} - \frac{1}{2}\boldsymbol{x}^T \boldsymbol{\Psi}^{-1}\boldsymbol{x}\right). \tag{1.27}$$

As shown in (1.25) and (1.27), in latent structure models with multivariate response variables, the random components can be described as follows. Let us set

$$\boldsymbol{\Psi} = \begin{pmatrix} a_1(\varphi_1) & & & \\ & a_2(\varphi_2) & 0 & \\ & 0 & \ddots & \\ & & & a_I(\varphi_I) \end{pmatrix},$$

$$\boldsymbol{\theta}^T = (\theta_1, \theta_2, \ldots, \theta_I), \boldsymbol{x}^T = (x_1, x_2, \ldots, x_I), 1 = (1, 1, \ldots, 1)^T.$$

Then, the random component can be expressed as

$$\begin{aligned} f(\boldsymbol{x}|\boldsymbol{\xi}) = \prod_{i=1}^{I} f_i(x_i|\boldsymbol{\xi}) &= \prod_{i=1}^{I} \exp\left(\frac{x_i\theta_i - b_i(\theta_i)}{a_i(\varphi_i)} + c_i(x_i, \varphi_i)\right) \\ &= \exp\left(\boldsymbol{x}^T \Psi^{-1}\boldsymbol{\theta} - 1^T \Psi^{-1}\boldsymbol{b}(\boldsymbol{\theta}) + \sum_{i=1}^{I} c_i(x_i, \varphi_i)\right). \end{aligned} \tag{1.28}$$

From the above discussion, by using appropriate linear predictors and link functions, latent structure models can be expressed by GLMs, for example, in the factor analysis model (1.21) and the latent class factor analysis model (1.22), the linear predictors are expressed by $\sum_{j=1}^{m}\lambda_{ij}\xi_j$ and the link functions are identity ones, and then, $\theta_i = \sum_{j=1}^{m}\lambda_{ij}\xi_j, i = 1, 2, \ldots, I$. Hence, the effects of latent variables on the manifest variables can be measured with the entropy coefficient of determination

(ECD) [4]. Let $f(\boldsymbol{x})$ and $g(\boldsymbol{\xi})$ be the marginal density or probability functions of manifest variable vector $\boldsymbol{X}$ and latent variable vector $\boldsymbol{\xi}$, respectively. Then, in the latent structure model (1.28), we have

$$\begin{aligned}\mathrm{KL}(\boldsymbol{X},\boldsymbol{\xi}) &= \int f(\boldsymbol{x}|\boldsymbol{\xi})g(\boldsymbol{\xi})\log\frac{f(\boldsymbol{x}|\boldsymbol{\xi})}{f(\boldsymbol{x})}d\boldsymbol{x}d\boldsymbol{\xi} + \int f(\boldsymbol{x})g(\boldsymbol{\xi})\log\frac{f(\boldsymbol{x})}{f(\boldsymbol{x}|\boldsymbol{\xi})}d\boldsymbol{x}d\boldsymbol{\xi} \\ &= \int (f(\boldsymbol{x}|\boldsymbol{\xi}) - f(\boldsymbol{x}))g(\boldsymbol{\xi})\log f(\boldsymbol{x}|\boldsymbol{\xi})d\boldsymbol{x}d\boldsymbol{\xi} = \mathrm{tr}\Psi^{-1}\mathrm{Cov}(\boldsymbol{\theta},\boldsymbol{X}).\end{aligned} \tag{1.29}$$

If the manifest and latent variables are discrete (categorical), the related integrals in (1.29) are substituted with appropriate summations. From the above KL information, the entropy coefficient of determination (ECD) is given by

$$\mathrm{ECD}(\boldsymbol{X},\boldsymbol{\xi}) = \frac{\mathrm{KL}(\boldsymbol{X},\boldsymbol{\xi})}{\mathrm{KL}(\boldsymbol{X},\boldsymbol{\xi})+1} = \frac{\mathrm{tr}\Psi^{-1}\mathrm{Cov}(\boldsymbol{\theta},\boldsymbol{X})}{\mathrm{tr}\Psi^{-1}\mathrm{Cov}(\boldsymbol{\theta},\boldsymbol{X})+1}. \tag{1.30}$$

The ECD expresses the explanatory or predictive power of the GLMs. Applying ECD to model (1.21), we have

$$\mathrm{ECD}(\boldsymbol{X},\boldsymbol{\xi}) = \frac{\sum_{i=1}^{I}\frac{\sum_{j=1}^{m}\lambda_{ij}^2}{\psi_i^2}}{\sum_{i=1}^{I}\frac{\sum_{j=1}^{m}\lambda_{ij}^2}{\psi_i^2}+1} = \frac{\sum_{i=1}^{I}\frac{R_i^2}{1-R_i^2}}{\sum_{i=1}^{I}\frac{R_i^2}{1-R_i^2}+1},$$

where R_i^2 are the coefficients of determination of predictors $\theta_i = \sum_{j=1}^{m}\lambda_{ij}\xi_j$ on the manifest variables $X_i, i = 1, 2, \ldots, I$ [5, 7]. Similarly, from model (1.22), we also get

$$\mathrm{ECD}(\boldsymbol{X},\boldsymbol{\xi}) = \frac{\sum_{i=1}^{I}\sum_{j=1}^{m}\lambda_{ij}^2\mathrm{Cov}(X_i,\xi_j)}{\sum_{i=1}^{I}\sum_{j=1}^{m}\lambda_{ij}^2\mathrm{Cov}(X_i,\xi_j)+1}.$$

Discussions of ECD in factor analysis and latent trait analysis are made in Eshima et al. [5] and Eshima [6]. In this book, ECD is used for measuring the predictive power of latent variables for manifest variables, and is also applied to make path analysis in latent class models.

Remark 1.4 In basic latent structure models treated in this chapter (1.23), since

$$\mathrm{KL}(X_i,\boldsymbol{\xi}) = \frac{\mathrm{Cov}(\theta_i, X_i)}{a(\varphi)}, i = 1, 2, \ldots, I,$$

from (1.29) we have

$$\mathrm{KL}(\boldsymbol{X}, \boldsymbol{\xi}) = \sum_{i=1}^{I} \mathrm{KL}(X_i, \boldsymbol{\xi}), = \sum_{i=1}^{I} \frac{\mathrm{Cov}(\theta_i, X_i)}{a_i(\varphi_i)}.$$

In models (1.21),

$$\mathrm{ECD}(X_i, \boldsymbol{\xi}) = \frac{\mathrm{KL}(X_i, \boldsymbol{\xi})}{\mathrm{KL}(X_i, \boldsymbol{\xi}) + 1} = R_i^2, i = 1, 2, \ldots, I.$$

1.7 The EM Algorithm and Latent Structure Models

The expectation–maximization (EM) algorithm [3] for the maximum likelihood (ML) estimation from incomplete data is reviewed in the latent structure model framework. The algorithm is a powerful tool for the ML estimation of latent structure models. In latent structure analysis, $(\boldsymbol{X}, \boldsymbol{\xi})$ and $\boldsymbol{X}$ are viewed as the complete and incomplete data, respectively. In the latent structure model, for parameter vector $\boldsymbol{\phi}$, the conditional density or probability function of $\boldsymbol{X}$ given $\boldsymbol{\xi}$, $f(\boldsymbol{x}|\boldsymbol{\xi})$, the marginal density or probability function of $\boldsymbol{X}$, $f(\boldsymbol{x})$, and that of $\boldsymbol{\xi}$, $g(\boldsymbol{\xi})$ are denoted by $f(\boldsymbol{x}|\boldsymbol{\xi})_{\boldsymbol{\phi}}$, $f(\boldsymbol{x})_{\boldsymbol{\phi}}$, and $g(\boldsymbol{\xi})_{\boldsymbol{\phi}}$, respectively. Let $f(\boldsymbol{x}, \boldsymbol{\xi})_{\boldsymbol{\phi}}$ be the joint density function of the complete data $(\boldsymbol{x}, \boldsymbol{\xi})$, then,

$$f(\boldsymbol{x}, \boldsymbol{\xi})_{\boldsymbol{\phi}} = f(\boldsymbol{x}|\boldsymbol{\xi})_{\boldsymbol{\phi}} g(\boldsymbol{\xi})_{\boldsymbol{\phi}},$$

and the log likelihood function of $\boldsymbol{\phi}$ based on incomplete data $\boldsymbol{X} = \boldsymbol{x}$ is expressed as

$$l(\boldsymbol{\phi}|\boldsymbol{x}) = \log f(\boldsymbol{x})_{\boldsymbol{\phi}}.$$

Let

$$Q\left(\boldsymbol{\phi}'|\boldsymbol{\phi}\right) = \mathrm{E}\left(\log f(\boldsymbol{x}, \boldsymbol{\xi})_{\boldsymbol{\phi}'}|\boldsymbol{x}, \boldsymbol{\phi}\right)$$

be the conditional expectation of $f(\boldsymbol{x}, \boldsymbol{\xi})_{\boldsymbol{\phi}'}$ given $\boldsymbol{X} = \boldsymbol{x}$ and parameter $\boldsymbol{\phi}$. The above conditional expectation is obtained by integrating with respect to latent variable vector $\boldsymbol{\xi}$. In order to get the ML estimates of the parameters $\boldsymbol{\phi}$ in latent structure models, $\widehat{\boldsymbol{\phi}}$, such that

$$l\left(\widehat{\boldsymbol{\phi}}\right) = \max_{\boldsymbol{\phi}} \log f(\boldsymbol{x})_{\boldsymbol{\phi}},$$

the EM algorithm is constituted of the following two steps:

(i) Expectation step (E-step)

For estimate ${}^{s+1}\boldsymbol{\phi}$ at the (s + 1) th step, compute the conditional expectation of $\log f(\boldsymbol{x}, \boldsymbol{\xi})_{\boldsymbol{\phi}}$ given the incomplete data $\boldsymbol{X} = \boldsymbol{x}$ and parameter ${}^{s}\boldsymbol{\phi}$:

$$Q\left(\boldsymbol{\phi}|{}^{s}\boldsymbol{\phi}\right) = \mathrm{E}\left(\log f(\boldsymbol{x}, \boldsymbol{\xi})_{\boldsymbol{\phi}}|\boldsymbol{x}, {}^{s}\boldsymbol{\phi}\right). \tag{1.31}$$

(ii) Maximization step (M-step)

Obtain $\boldsymbol{\phi}^{s+1}$ such that

$$Q\left({}^{s+1}\boldsymbol{\phi}|{}^{s}\boldsymbol{\phi}\right) = \max_{\boldsymbol{\phi}} Q\left(\boldsymbol{\phi}|{}^{s}\boldsymbol{\phi}\right). \tag{1.32}$$

■

By using the above iterative procedure, the ML estimates of latent structure models can be obtained. If there exists a sufficient statistic $t(\boldsymbol{x}, \boldsymbol{\xi})$ for parameter vector $\boldsymbol{\phi}$ such that

$$f(\boldsymbol{x}, \boldsymbol{\xi}|\boldsymbol{\phi}) = b(\boldsymbol{x}, \boldsymbol{\xi})\frac{\exp\left(\boldsymbol{\phi}t(\boldsymbol{x}, \boldsymbol{\xi})^{T}\right)}{s(\boldsymbol{\phi})}, \tag{1.33}$$

the EM algorithm is simplified as follows:

(i) E-step

Compute

$${}^{s+1}\boldsymbol{t} = E\left(t(\boldsymbol{x}, \boldsymbol{\xi})|\boldsymbol{x}, {}^{s}\boldsymbol{\phi}\right). \tag{1.34}$$

(ii) M-step

Obtain $\boldsymbol{\phi}^{p+1}$ from the following equation:

$${}^{s+1}\boldsymbol{t} = E(t(\boldsymbol{X}, \boldsymbol{\xi})|\boldsymbol{\phi}). \tag{1.35}$$

1.8 Discussion

Basic latent structure models, i.e., the latent class model, latent trait model, latent profile model, and factor analysis model, are overviewed in this chapter. These models are based on, what we call, the assumption of local independence, that is, the manifest variables are statistically independent, given latent variables. These models can be expressed in a GLM framework, and the multivariate formulation of latent structure models is also given in (1.28). Studies of latent structure models through a GLM framework will be important to grasp the models in a general way and to apply them in various research domains, and it may lead to the construction of new latent structure models. It is expected that new latent structure models are designed in the applications. The EM algorithm is a useful tool to perform the ML estimation of latent structure models, and a brief review of the method is also given in this chapter. In the following chapters, the EM algorithm is used to estimate the model parameters.

References

1. Bartholomew, D. J. (1987). *Latent variable models and factor analysis*. Charles & Griffin.
2. Dayton, M., & Macready, G. B. (1976). A probabilistic model for validation of behavioral hierarchies. *Psychometrika, 41*, 190–204.
3. Dempster, A. P., Laird, N. M., & Rubin, D. B. (1977). Maximum likelihood from incomplete data via the EM algorithm (with discussion). *Journal of the Royal Statistical Society, B, 39*, 1–38.
4. Eshima, N., & Tabata, M. (2010). Entropy coefficient of determination for generalized linear models. *Computational Statistics and Data Analysis, 54*, 1381–1389.
5. Eshima, N., Tabata, M., & Borroni, C. G. (2018). An entropy-based approach for measuring factor contributions in factor analysis models. *Entropy, 20*, 634.
6. Eshima, N. (2020). *Statistical data analysis and entropy*. Springer Nature.
7. Eshima, N., Borroni, C. G., Tabata, M., & Kurosawa, T. (2021). An entropy-based tool to help the interpretation of common-factor spaces in factor analysis. *Entropy, 23*, 140. https://doi.org/10.3390/e23020140-24
8. Everitt, B. S. (1984). *An introduction to latent variable models*. Chapman & Hall.
9. Gibson, W. A. (1959). Three multivariate models: Factor analysis, latent structure analysis, and latent profile analysis, *Psychometrika 24*, 229–252.
10. Lazarsfeld, P. F., & Henry, N. M. (1968). *Latent structure analysis*. Houghton Mifflin.
11. Lazarsfeld, P. F. (1959). *Latent structure analysis, psychology: A study of a science,* Koch, S. ed. McGrowHill: New York.
12. Lazarsfeld, P. F. (1950). The logical and mathematical foundation of latent structure analysis. In Soufer, S. A., Guttman, L., & others (Eds.), *Measurement and prediction: Studies in social psychology I World War II* (Vol. 4). Prenceton University Press.
13. Lord, F. M., & Novick, M. R. (1968). *Statistical theories of mental test scores*. Addison-Wesley.
14. Lord, F. M. (1952). *A theory of test scores* (Psychometric Monograph, No. 7), Richmond VA: Psychometric Corporation.
15. Magidson, J., & Vermunt, J. K. (2001). Latent class factor and cluster models: Bi-plots, and related graphical displays. *Sociological Methodology, 31*, 223–264.
16. McCullagh, P., & Nelder, J. A. (1989). *Generalized linear models* (2nd ed.). London.
17. Nelder, J. A., & Wedderburn, R. W. M. (1972). Generalized linear model. *Journal of the Royal Statistical Society A, 135*, 370–384.
18. Rasch, G. (1960). *Probabilistic model for some intelligence and attainment tests*. Danish Institute for Educational Research.
19. Samejima, F. (1973). A method of estimating item characteristic functions using the maximum likelihood estimate od ability. *Psychometrika, 38*, 163–191.
20. Samejima, F. (1974). Normal ogive model on the continuous response level in the multidimensional latent space. *Psychometrika, 39*, 111–121.
21. Spearman, S. (1904). "General-intelligence", objectively determined and measured. *American Journal of Psychology, 15*, 201–293.
22. Thurstone, L. L. (1935). *Vector of mind*: *Multiple factor analysis for the isolation of primary traits*. Chicago, IL, USA: The University of Chicago Press.

Chapter 2
Latent Class Cluster Analysis

2.1 Introduction

In behavioral sciences, there are many cases where we can assume that human behaviors and responses depend on latent concepts, which are not directly observed. In such cases, it is significant to elucidate the latent factors to affect and cause human behaviors and responses. For this objective, latent class analysis was proposed by Lazarsfeld [11] to explore discrete (categorical) latent factors that explain the relationships among responses to items under studies. The responses to the items are treated by manifest variables and the factors by latent variables. By use of models with manifest and latent variables, it is possible to analyze the phenomena concerned. A general latent class model is expressed with (1.2) and (1.3), and for binary manifest variables the model is expressed by (1.4). These equations are called accounting equations. The parameters in models (1.2) and (1.4) are manifest probabilities $P(\boldsymbol{X} = \boldsymbol{x})$ and latent probabilities v_a, $P(X_i = x_i|a)$, and π_{ai}. Although the manifest probabilities can be estimated directly by the relative frequencies of responses $\boldsymbol{X} = \boldsymbol{x}$ as consistent estimates, the latent probabilities cannot be estimated easily. In the early stages of the development of latent class analysis, the efforts for the studies were concentrated on the parameter estimation by solving the accounting equations, for example, Green [10], Anderson [1], Gibson [6, 7], Madansky [13], and so on; however, these studies are now only in the study history. As increasing computer efficiency, methods for the ML estimation were widely applied; however, it was critical to obtain proper estimates of the latent probabilities, that is, the estimates have to be between 0 and 1. The usual ML estimation methods often derived the improper solutions in real data analyses, in which improper solutions imply the latent probability estimates that are outside of interval [0, 1]. To overcome the problem, two ways for the ML estimation were proposed. One is a proportional fitting method by Goodman [8, 9], which is included in the EM algorithm for the ML estimation [3]. The second is a method in which a parameter transformation is employed to deal with the ML estimation by a direct use of the Newton–Raphson algorithm [5]. Although the convergence rate of

N. Eshima, *An Introduction to Latent Class Analysis*, Behaviormetrics: Quantitative Approaches to Human Behavior 14,
https://doi.org/10.1007/978-981-19-0972-6_2

Goodman's method is slow, the method is simple and flexible to apply in real data analysis. In this sense, Goodman's contribution to the development of latent class analysis is great.

This chapter consists of seven sections including this section. In Sect. 2.2, Goodman's method for the ML estimation of the latent class model is derived from the EM algorithm and the property of the algorithm is discussed. Section 2.3 applies the ML estimation algorithm to practical data analyses. In Sect. 2.4, in order to measure the goodness-of-fit of latent class models, the entropy coefficient of determination [4] is used. Section 2.5 considers two methods for comparing latent classes are discussed and the methods are illustrated by using numerical examples. In Sect. 2.6, a method for the ML estimation of the latent profile model is constructed according to the EM algorithm, and a numerical example is also given to demonstrate the method. Finally, Sect. 2.7 provides a discussion on the latent class analysis presented in this chapter and a perspective of the analysis leading to further studies in the future.

2.2 The ML Estimation of Parameters in the Latent Class Model

In latent class model (1.2) with constraints (1.3), the complete data would be obtained as responses to I items $\boldsymbol{X} = (X_1, X_2, \ldots, X_I)^T$ in latent classes a. Let $n(x_1, x_2, \ldots, x_I)$ and $n(x_1, x_2, \ldots, x_I, a)$ be the numbers of observations with response $\boldsymbol{x} = (x_1, x_2, \ldots, x_I)^T$ and those in latent class a, respectively, and let $\boldsymbol{\phi} = ((v_a), (P(X_i = x_i|a)))$ be the parameter row vector. Concerning numbers of observations $n(\boldsymbol{x})$ and $n(\boldsymbol{x}, a)$, it follows that

$$n(\boldsymbol{x}) = \sum_{\alpha=1}^{A} n(\boldsymbol{x}, a).$$

Since the probability of $\boldsymbol{X} = \boldsymbol{x}$ is expressed by

$$P(\boldsymbol{X} = \boldsymbol{x}) = \sum_{\alpha=1}^{A} v_a \prod_{i=1}^{I} P(X_i = x_i|a), \tag{2.1}$$

the log likelihood function of parameter vector $\boldsymbol{\phi}$ based on incomplete data $(n(\boldsymbol{x}))$ is given by

$$l(\boldsymbol{\phi}|(n(\boldsymbol{x}))) = \sum_{\boldsymbol{x}} n(\boldsymbol{x}) \log\left(\sum_{\alpha=1}^{A} v_a \prod_{i=1}^{I} P(X_i = x_i|a)\right), \tag{2.2}$$

where the summation in the above formula $\sum_{\boldsymbol{x}}$ is made over all response patterns $\boldsymbol{x} = (x_1, x_2, \ldots, x_I)$. Since the direct maximization of log likelihood function (2.2) with respect to $\boldsymbol{\phi}$ is very complicated, the EM algorithm is employed. Given that complete data, statistics $\boldsymbol{t}(\boldsymbol{x}, a) = (n(\boldsymbol{x}, a))$ in (1.33) are obtained as a sufficient statistic vector for parameters $\boldsymbol{\phi}$, and we have the log likelihood function of parameter vector $\boldsymbol{\phi}$ as follows:

$$\begin{aligned} l(\phi|(n(\boldsymbol{x}, a))) &= \sum_{\alpha=1}^{A}\sum_{\boldsymbol{x}} n(\boldsymbol{x}, a)\log\left(v_a \prod_{i=1}^{I} P(X_i = x_i|a)\right) \\ &= \sum_{\alpha=1}^{A}\sum_{\boldsymbol{x}} n(\boldsymbol{x}, a)\left(\log v_a + \sum_{i=1}^{I} \log P(X_i = x_i|a)\right). \end{aligned} \tag{2.3}$$

In this sense, sufficient statistic vector $\boldsymbol{t}(\boldsymbol{x}, a) = (n(\boldsymbol{x}, a))$ is viewed as the complete data. Let ${}^{s}\boldsymbol{\phi} = (({}^{s}v_a), ({}^{s}P(X_i = x_i|a)))$ be the estimate of $\boldsymbol{\phi}$ at the sth iteration in the EM algorithm. Then, from (1.34) and (1.35), the E- and M-steps are formulated as follows.

The EM algorithm for model (1.2) with constraints (1.3).

(i) E-step

Let

$${}^{s+1}\boldsymbol{t}(\boldsymbol{x}, a) = \left({}^{s+1}n(\boldsymbol{x}, a)\right)$$

be the conditional expectation of the sufficient statistic for the given incomplete (observed) data $\boldsymbol{x}$ and parameters ${}^{s}\boldsymbol{\phi}$. From (1.34), we have

$${}^{s+1}n(\boldsymbol{x}, a) = n(\boldsymbol{x})\frac{{}^{s}v_a \prod_{i=1}^{I} {}^{s}P(X_i = x_i|a)}{\sum_{b=1}^{A} {}^{s}v_b \prod_{i=1}^{I} {}^{s}P(X_i = x_i|b)}. \tag{2.4}$$

From the above results, we can get ${}^{s+1}\boldsymbol{t}(\boldsymbol{x}, a) = \left({}^{s+1}n(\boldsymbol{x}, a)\right)$.

(ii) M-step

Let Ω be the sample space of the manifest variable vector $\boldsymbol{X}$; let $\Omega(X_i = x_i)$ be the sample subspaces of $\boldsymbol{X}$ for given $X_i = x_i, i = 1, 2, \ldots, I$; let Ω_i be the sample space of manifest variables $X_i, i = 1, 2, \ldots, I$; and let $N = \sum_{\boldsymbol{x}} n(\boldsymbol{x})$. From (1.35), we have

$${}^{s+1}n(\boldsymbol{x}, a) = Nv_a \prod_{i=1}^{I} P(X_i = x_i|a), \boldsymbol{x} \in \Omega, a = 1, 2, \ldots, A. \tag{2.5}$$

The following constraints hold true:

$$\sum_{x\in\Omega}\prod_{i=1}^{I} P(X_i = x_i|a) = 1, a = 1, 2, \ldots, A,$$

and

$$\sum_{x\in\Omega(X_i=x_i)}\prod_{j=1}^{I} P\left(X_j = x_j|a\right) = P(X_i = x_i|a), i = 1, 2, \ldots, I; a = 1, 2, \ldots, A,$$

where $\sum_{x\in\Omega}$ is the summation over all response patterns $\boldsymbol{X} = \boldsymbol{x}$, and $\sum_{x\in\Omega(X_i=x_i)}$, the summation over all response patterns $\boldsymbol{X} = \boldsymbol{x}$ for given $X_i = x_i$. Solving equations in (2.5), with respect to parameters v_a, $P(X_i = x_i|a)$, $i = 1, 2, \ldots, I$, under constraints (1.3), it follows that

$$\begin{cases} {}^{s+1}v_a = \frac{1}{N}\sum\limits_{x\in\Omega}^{s+1} n(\boldsymbol{x}, a),\ a = 1, 2, \ldots, A; \\ {}^{s+1}P(X_i = x_i|a) = \dfrac{1}{N\cdot{}^{s+1}v_a}\sum\limits_{x\in\Omega(X_i=x_i)} {}^{s+1}n(\boldsymbol{x}, a), \\ x_i \in \Omega_i,\ i = 1, 2, \ldots, I; a = 1, 2, \ldots, A. \end{cases} \tag{2.6}$$

The above algorithm is the same as a proportional fitting method by Goodman [8, 9], and the ML estimates of the parameters $\hat{v}_a$ and $\widehat{P}(X_i = x_i|a)$ can be obtained as the convergence values of the above estimates, ${}^{\infty}v_a$ and ${}^{\infty}P(X_i = x_i|a)$. From (2.6), it is seen that for any integer s,

$$\begin{cases} 0 \le {}^{s}v_a \le 1, a = 1, 2, \ldots, A; \\ 0 \le {}^{s}P(X_i = x_i|a) \le 1, x_i \in \Omega_i, i = 1, 2, \ldots, I; a = 1, 2, \ldots, A. \end{cases}$$

Thus, if the above algorithm converges, the estimates are proper for the likelihood function, and satisfy the following equations:

$$\begin{cases} \frac{\partial}{\partial v_a} l(\boldsymbol{\phi}|(n(\boldsymbol{x}))) = 0, a = 1, 2, \ldots, A; \\ \frac{\partial}{\partial P(X_i=x_i|a)} l(\boldsymbol{\phi}|(n(\boldsymbol{x}))) = 0, x_i \in \Omega_i, i = 1, 2, \ldots, I; a = 1, 2, \ldots, A. \end{cases}$$

For observed data set $\{n(\boldsymbol{x})\}$, the goodness-of-fit test of a latent class model to the data can be carried out with the following log likelihood ratio test statistic:

$$G^2 = 2\sum_{x} n(\boldsymbol{x})\left\{\log\frac{n(\boldsymbol{x})}{N} - \log\left(\sum_{\alpha=1}^{A}\hat{v}_a\prod_{i=1}^{I}\widehat{P}(X_i = x_i|a)\right)\right\}. \tag{2.7}$$

For sufficiently large sample size N, the above statistic is asymptotically χ^2-distributed with degrees of freedom, the number of manifest parameters $P(\boldsymbol{X} = \boldsymbol{x})$

minus latent parameters v_a and $P(X_i = x_i|a)$, that is, from (1.5) we have

$$\prod_{i=1}^{I} K_i - 1 - (A-1) - A\sum_{i=1}^{I}(K_i - 1) = \prod_{i=1}^{I} K_i - A\left\{\sum_{i=1}^{I} K_i - (I-1)\right\}. \quad (2.8)$$

After estimating a latent class model under study, the interpretation of latent classes is made by considering the sets of the estimated latent response probabilities $\left\{\widehat{P}(X_i = x_i|a), i = 1, 2, \ldots, I\right\}, a = 1, 2, \ldots, A$. In addition to the interpretation, it is significant to assess the manifest response vectors $\boldsymbol{x} = (x_1, x_2, \ldots, x_I)$ with respect to latent classes, that is, to assign individuals with the manifest responses to the extracted latent classes. The best way to assign them to the latent classes is made with the maximum posterior probabilities, that is, if

$$P(a_0|(x_1, x_2, \ldots, x_I)) = \max_a \frac{v_a \prod_{i=1}^{I} P(X_i = x_i|a)}{\sum_{b=1}^{A} v_b \prod_{i=1}^{I} P(X_i = x_i|b)}, \quad (2.9)$$

an individual with response $\boldsymbol{x} = (x_1, x_2, \ldots, x_I)$ is evaluated as a member in latent class a_0.

Remark 2.1 The EM algorithm for the latent class model, mentioned above, has been constructed by using E-step (1.34) and M-step (1.35). The algorithm can also be directly derived with (1.31) and (1.32). The process is given as follows:

(i) E-step

$$\begin{aligned} Q(\phi|^s\phi) &= \mathrm{E}\left(\sum_{\alpha=1}^{A}\sum_{\boldsymbol{x}} n(\boldsymbol{x}, a)\left(\log v_a + \sum_{i=1}^{I}\log P(X_i = x_i|a)\right)|\boldsymbol{x}, {}^s\phi\right) \\ &= \sum_{\alpha=1}^{A}\sum_{\boldsymbol{x}} \mathrm{E}\left(n(\boldsymbol{x}, a)|\boldsymbol{x}, {}^s\phi\right)\left(\log v_a + \sum_{i=1}^{I}\log P(X_i = x_i|a)\right) \\ &= \sum_{\alpha=1}^{A}\sum_{\boldsymbol{x}}^{s+1} n(\boldsymbol{x}, a)\left(\log v_a + \sum_{i=1}^{I}\log P(X_i = x_i|a)\right), \end{aligned}$$

where ${}^{s+1}n(\boldsymbol{x}, a)$ are given in (2.4).

(ii) M-step

Considering the constraints in (1.3), let λ and $\mu_{ai}, a = 1, 2, \ldots, A; i = 1, 2, .., .I$ be Lagrange multipliers. Then, the Lagrange function is given by

$$L(\phi|^s\phi) = Q(\phi|^s\phi) - \lambda\sum_{a=1}^{A} v_a - \sum_{a=1}^{A}\sum_{i=1}^{I}\mu_i\sum_{x_i} P(X_i = x_i|a)$$

$$= \sum_{\alpha=1}^{A} \sum_{\boldsymbol{x}}^{s+1} n(\boldsymbol{x}, a)\left(\log v_a + \sum_{i=1}^{I} \log P(X_i = x_i|a)\right) - \lambda \sum_{a=1}^{A} v_a$$
$$- \sum_{a=1}^{A} \sum_{i=1}^{I} \mu_{ai} \sum_{x_i} P(X_i = x_i|a).$$

Differentiating the Lagrange function with respect to v_a, we have

$$\frac{\partial}{\partial v_a} L(\boldsymbol{\phi}|^s\boldsymbol{\phi}) = \frac{\partial}{\partial v_a} Q(\boldsymbol{\phi}|^s\boldsymbol{\phi}) - \lambda = \frac{1}{v_a} \sum_{\boldsymbol{x}} {}^{s+1}n(\boldsymbol{x}, a) - \lambda = 0, a = 1, 2, \ldots, A.$$

From the above equations, it follows that

$$v_a \lambda = \sum_{\boldsymbol{x}} {}^{s+1}n(\boldsymbol{x}, a), a = 1, 2, \ldots, A.$$

Summing up both sides of the above equations for $a = 1, 2, \ldots, A$, we obtain

$$\lambda = \sum_{a=1}^{A} \sum_{\boldsymbol{x}} {}^{s+1}n(\boldsymbol{x}, a) = N.$$

Hence, the $s + 1$ th estimates of v_a are derived as follows:

$${}^{s+1}v_a = \frac{\sum_{\boldsymbol{x}} {}^{s+1}n(\boldsymbol{x}, a)}{\lambda} = \frac{\sum_{\boldsymbol{x}} {}^{s+1}n(\boldsymbol{x}, a)}{N}, a = 1, 2, \ldots, A. \tag{2.10}$$

Similarly, differentiating the Lagrange function with respect to $P(X_i = x_i|a)$, we have

$$\frac{\partial L(\boldsymbol{\phi}|^s\boldsymbol{\phi})}{\partial P(X_i = x_i|a)} = \frac{\partial Q(\boldsymbol{\phi}|^s\boldsymbol{\phi})}{\partial P(X_i = x_i|a)} - \mu_{ai} = \frac{\sum_{\boldsymbol{x} \in \Omega(X_i = x_i)} {}^{s+1}n(\boldsymbol{x}, a)}{P(X_i = x_i|a)} - \mu_{ai} = 0,$$
$$i = 1, 2, \ldots, I; a = 1, 2, \ldots, A.$$

From this, it follows that

$$P(X_i = x_i|a)\mu_{ai} = \sum_{\boldsymbol{x} \in \Omega(X_i = x_i)} {}^{s+1}n(\boldsymbol{x}, a), i = 1, 2, \ldots, I; a = 1, 2, \ldots, A.$$

Summing up both sides of the above equations with respect to x_i, we get

$$\mu_{ai} = \sum_{x_i=1}^{K_i} \sum_{\boldsymbol{x} \in \Omega(X_i = x_i)} {}^{s+1}n(\boldsymbol{x}, a) = \sum_{\boldsymbol{x}} {}^{s+1}n(\boldsymbol{x}, a), i = 1, 2, \ldots, I; a = 1, 2, \ldots, A.$$

Thus, (2.6) is obtained as follows:

$$ {}^{s+1}P(X_i = x_i|a) = \frac{\sum_{\boldsymbol{x}\in\Omega(X_i=x_i)} {}^{s+1}n(\boldsymbol{x}, a)}{\sum_{\boldsymbol{x}} {}^{s+1}n(\boldsymbol{x}, a)}, i = 1, 2, \ldots, I; a = 1, 2, \ldots, A. \tag{2.11} $$

2.3 Examples

Table 2.1 shows the data from respondents to questionnaire items on role conflict [20], and the respondents are cross-classified with respect to whether they tend toward universalistic values "1" or particularistic values "0" when confronted by each of four different situations of role conflict [18]. Assuming A latent classes in the population, latent class model (2.1) is applied. Let $X_i, i = 1, 2, 3, 4$ be the responses to the four situations. According to the condition of model identification, the formula in (2.8) have to be positive, so we have

$$ 2^4 - A\{2 \times 4 - (4 - 1)\} = 16 - 5A > 0. $$

From the above inequality, the number of latent classes has to be less than and equal to $\frac{16}{5}$. Assuming three latent classes, with which the latent class model is denoted by $M(3)$, the EM algorithm with (2.4) and (2.6) is carried out. The ML estimates of the parameters are illustrated in Table 2.2, and the following inequalities hold:

$$ \widehat{P}(X_i = 1|1) < \widehat{P}(X_i = 1|2) < \widehat{P}(X_i = 1|3), i = 1, 2, 3, 4. $$

From the above results, the extracted latent classes 1, 2, and 3 in Table 2.2 can be interpreted as ordered latent classes, "low", "medium", and "high", in the universalistic attitude in the role of conflict. The latent class model with three latent classes for

Table 2.1 Data of responses in four different situations of role conflict

Response pattern	Frequency	Response pattern	Frequency
0 0 0 0	20	0 0 0 1	2
1 0 0 0	38	1 0 0 1	7
0 1 0 0	6	0 1 0 1	1
1 1 0 0	25	1 1 0 1	6
0 0 1 0	9	0 0 1 1	2
1 0 1 0	24	1 0 1 1	6
0 1 1 0	4	0 1 1 1	1
1 1 1 0	23	1 1 1 1	42

Source Stouffer and Toby [20], Goodman [8]

Table 2.2 The estimates of the parameters for a latent class model with three latent classes (Stouffer-Toby data in Table 2.1)

Latent class	Proportion	Latent positive item response probability			
		X_1	X_2	X_3	X_4
1	0.220	0.005	0.032	0.024	0.137
2	0.672	0.194	0.573	0.593	0.830
3	0.108	0.715	1.000	0.759	0.943

[a] The log likelihood ratio test statistic (2.6) is calculated as $G^2(3) = 0.387(df = 1, P = 0.534)$

Table 2.3 The estimates of the parameters for a latent class model with two latent classes (Stouffer-Toby data in Table 2.1)

Latent class	Class proportion	Latent positive item response probability			
		X_1	X_2	X_3	X_4
1	0.279	0.007	0.060	0.073	0.231
2	0.721	0.286	0.670	0.646	0.868

[a] $G^2(2) = 2.720(df = 6, P = 0.843)$

four binary items has only one degree of freedom left for the test of goodness-of-fit to the data, so a latent class model with two latent classes $M(2)$ is estimated and the results are shown in Table 2.3. In this case, two ordered latent classes, "low" and "high" in the universalistic attitude, are extracted. The goodness-of-fit of both models to the data is good. In order to compare the two models, the relative goodness-of-fit of $M(2)$ to $M(3)$ can be assessed by

$$G^2(2) - G^2(3) = 2.720 - 0.387 = 2.333, df = 5, P = 0.801.$$

From this, $M(2)$ is better than $M(3)$ to explain the present response behavior. Stouffer and Toby [20] observed the data in Table 2.1 to order the respondents in a latent continuum with respect to the relative priority of personal and impersonal considerations in social obligations. In this sense, it is significant to have obtained the ordered latent classes in the present latent class analysis. According to posterior probabilities (2.9), the assessment results of respondents with manifest responses $\boldsymbol{x} = (x_1, x_2, x_3, x_4)^T$ for $M(2)$ and $M(3)$ are demonstrated in Table 2.4. Both results are almost the same. As shown in this data analysis, we can assess the respondents with their response patterns $\boldsymbol{x} = (x_1, x_2, x_3, x_4)$, not simple total of the responses to test items $\sum_{i=1}^{4} x_i$ (Table 2.4).

Table 2.5 illustrates test data on creative ability in machine design [15]. Engineers are cross-classified with respect to their dichotomized scores, that is, above the subtest mean (1) or below (0), obtained on each of four subtests that measured creative abilities in machine design [18]. If we can assume a one-dimensional latent continuum with respect to the creative ability, it may be reasonable to expect to

Table 2.4 Assignment of the manifest responses to the extracted latent classes (Data in Table 2.1)

Response pattern	$M(2)$ latent class	$M(3)$ latent class	Response pattern	$M(2)$ latent class	$M(3)$ latent class
0 0 0 0	1	1	0 0 0 1	1	2
1 0 0 0	2	2	1 0 0 1	2	2
0 1 0 0	2	2	0 1 0 1	2	2
1 1 0 0	2	2	1 1 0 1	2	2
0 0 1 0	1	2	0 0 1 1	2	2
1 0 1 0	2	2	1 0 1 1	2	2
0 1 1 0	2	2	0 1 1 1	2	2
1 1 1 0	2	2	1 1 1 1	2	3

Table 2.5 Data on creative ability in machine design (McHugh's data)

Response pattern	Frequency	Response pattern	Frequency
0 0 0 0	23	0 0 0 1	5
1 0 0 0	6	1 0 0 1	3
0 1 0 0	8	0 1 0 1	2
1 1 0 0	9	1 1 0 1	3
0 0 1 0	5	0 0 1 1	14
1 0 1 0	2	1 0 1 1	4
0 1 1 0	3	0 1 1 1	8
1 1 1 0	8	1 1 1 1	34

Source McHugh [15], Proctor (1970)

derive ordered latent classes in latent class analysis as in the analysis of Stouffer-Toby's data. First, for three latent classes, we have the results of latent class analysis shown in Table 2.6. The goodness-of-fit of the model to the data set is bad, since we get $G^2(3) = 4.708(df = 1, P = 0.030)$. Similarly, for a latent class model with two latent classes, the goodness-of-fit of the model to the data set is also bad, that is, $G^2(2) = 25.203(df = 6, P = 0.000)$. From the results, it is not appropriate to apply the latent class cluster analysis to the data set. For each of the four different

Table 2.6 The estimates of the parameters for a latent class model with three latent classes (data in Table 2.5)

Latent class	Class proportion	Latent positive item response probability			
		X_1	X_2	X_3	X_4
1	0.198	0.239	0.000	0.808	0.803
2	0.398	0.324	0.360	0.089	0.111
3	0.404	0.810	1.000	0.926	0.810

[a] $G^2(3) = 4.708(df = 1, P = 0.030)$

subtests, it may be needed to assume a particular skill to obtain scores above the mean, where the four skills cannot be ordered with respect to difficulty for obtaining them. Assuming the particular skills for solving the subtests, a confirmatory latent class analysis of the data is carried out in Chap. 4.

The third data (Table 2.7) were obtained from noncommissioned officers to items on attitude toward the Army [18]. The respondents were cross-classified with respect to their dichotomous responses, which were made according to dichotomized responses "1" as "favorable" and "0" "unfavorable" toward the Army for each of the four different items on general attitude toward the Army. If there exists a latent continuum with respect to the attitude, we can assume ordered latent classes as in the first data (Table 2.1). The estimated latent class models with three and two latent classes are given in Tables 2.8 and 2.9, respectively. As shown in the results of the test of the goodness-of-fit of the models, the degrees of the models are fair. As shown in the tables, the estimated latent classes can be ordered, because, for example, in Table 2.8, the following inequalities hold:

$$\widehat{P}(X_i = 1|1) < \widehat{P}(X_i = 1|2) < \widehat{P}(X_i = 1|3), i = 1, 2, 3, 4.$$

Hence, the extracted latent classes 1–3 can be interpreted as "low", "medium", and "high" groups in favorable attitude toward the Army, respectively. Comparing

Table 2.7 Data on attitude toward the Army (Lazarsfeld-Stouffer's data)

Response pattern	Frequency	Response pattern	Frequency
0 0 0 0	75	0 0 0 1	69
1 0 0 0	3	1 0 0 1	16
0 1 0 0	42	0 1 0 1	60
1 1 0 0	10	1 1 0 1	25
0 0 1 0	55	0 0 1 1	96
1 0 1 0	8	1 0 1 1	52
0 1 1 0	45	0 1 1 1	199
1 1 1 0	16	1 1 1 1	229

Source Price et al. [18]

Table 2.8 The estimates of the parameters for a latent class model with three latent classes (Lazarsfeld-Stouffer's data in Table 2.7)

Latent class	Class proportion	Latent positive item response probability			
		X_1	X_2	X_3	X_4
1	0.260	0.000	0.296	0.386	0.406
2	0.427	0.374	0.641	0.672	0.768
3	0.313	0.637	0.880	1.000	1.000

[a] $G^2(3) = 1.787(df = 1,\ P = 0.181)$

Table 2.9 The estimates of the parameters for a latent class model with two latent classes (Lazarsfeld-Stouffer's data in Table 2.7)

Latent class	Class proportion	Latent positive item response probability			
		X_1	X_2	X_3	X_4
1	0.445	0.093	0.386	0.442	0.499
2	0.555	0.572	0.818	0.906	0.944

[a] $G^2(2) = 8.523(df = 6, P = 0.202)$

the models, since the relative goodness-of-fit of $M(2)$ is

$$G^2(2) - G^2(3) = 8.523 - 1.787 = 6.736, df = 5, P = 0.241,$$

model $M(2)$ is better than $M(3)$. The data are treated in Chapter 3 again, assuming ordered latent classes.

Comparing Tables 2.2, 2.6, and 2.8, each of Tables 2.2 and 2.8 shows three ordered latent classes; however, the estimated latent classes in Table 2.6 are not consistently ordered with respect to positive response probabilities for four test items. It can be thought the universalistic attitude in the role conflict in Stouffer-Toby's data and the favorable attitude toward the Army are one-dimensional, but in machine design, latent classes may not be assessed one-dimensionally.

2.4 Measuring Goodness-of-Fit of Latent Class Models

As in the ordinary linear regression analysis, it is meaningful to evaluate the predictive power or goodness-of-fit of the latent class model. According to a GLM framework of latent structure models (Chap. 1, Sect. 1.6), the KL information (1.29) is applied to the latent class model (2.1). In order to facilitate the discussion, the application is made for latent class models with binary response variables (1.4). According to the assumption of local independence, from (1.24), (1.25), and (1.29), we have

$$\mathrm{KL}(\boldsymbol{X}, \boldsymbol{\xi}) = \sum_{i=1}^{I} \mathrm{KL}(X_i, \boldsymbol{\xi}) = \sum_{i=1}^{I} \mathrm{Cov}(\theta_i, X_i). \tag{2.12}$$

It means that "the variation of manifest variable vector $\boldsymbol{X}$ in entropy" explained by latent classes is decomposed into those of manifest variables X_i. Since from (1.25)

$$E(X_i) = \sum_{\alpha=1}^{A} v_a \pi_{ai} (= \pi_i),$$

Table 2.10 Assessment of latent class model $M(3)$ in Table 2.2

Manifest variable	X_1	X_2	X_3	X_4	(X_1, X_2, X_3, X_4)
KL	0.298	0.564	0.547	0.480	1.888
ECD	0.229	0.361	0.354	0.324	0.654

Table 2.11 Assessment of latent class model $M(2)$ in Table 2.3

Manifest variable	X_1	X_2	X_3	X_4	(X_1, X_2, X_3, X_4)
KL	0.229	0.425	0.361	0.395	1.410
ECD	0.186	0.298	0.265	0.283	0.585

$$E(\theta_i) = \sum_{\alpha=1}^{A} v_a \theta_{ai} = \sum_{\alpha=1}^{A} v_a \log \frac{\pi_{ai}}{1 - \pi_{ai}},$$

and we have

$$\mathrm{KL}(X_i, \boldsymbol{\xi}) = \mathrm{Cov}(\theta_i, X_i) = \sum_{a=1}^{A} v_a \left(\pi_{ai} - \pi_i\right) \log \frac{\pi_{ai}}{1 - \pi_{ai}}, i = 1, 2, \ldots, I. \quad (2.13)$$

Increasing the above information, stronger is the association between manifest variables X_i and latent variable (class) $\boldsymbol{\xi}$. ECD in GLMs corresponds to the coefficient of determination R^2 in the ordinary linear regression models. The above discussion is applied to Table 2.2, and we calculate the KL information and ECDs (Table 2.10). 65.4% of the variation of response variable vector $\boldsymbol{X} = (X_1, X_2, X_3, X_4)^T$ in entropy is explained by the latent classes. According to the KL information criterion, the association of manifest variable X_2 with the latent variable is the strongest among the manifest variables. For latent class model $M(2)$ in Table 2.3, the same assessment is made and the results are illustrated in Table 2.11. The results are similar to those for $M(3)$ in Table 2.10.

2.5 Comparison of Latent Classes

In latent class model (2.1), the latent classes are interpreted with the latent response probabilities to items, $\{P(X_i = x_i|a), i = 1, 2, \ldots, I\}, \alpha = 1, 2, \ldots, A$. When there are two latent classes in a population, we can always say one latent class is higher than the other one in a concept or latent trait. However, where the number of latent classes is greater than two, we cannot easily assess and compare the latent classes without latent concepts or traits, and so it is meaningful to make methods for comparing latent classes in latent class cluster analysis. In this section, first, a technique similar to canonical analysis is employed to make a latent space to compare the latent classes, that is, to construct a latent space for locating the latent classes. We discuss the case where the manifest variables X_i are binary. Let

$$\pi_{ai} = P(X_i = 1|a), i = 1, 2, \ldots, I; a = 1, 2, \ldots, A$$

and

$$\pi_i = P(X_i = 1)\left(= \sum_{a=1}^{A} v_a \pi_{ai} \right).$$

For manifest responses $\boldsymbol{X} = (X_1, X_2, \ldots, X_I)^T$, the following score is given:

$$T = \sum_{i=1}^{I} \{c_{i0}(1 - X_i) + c_{i1}X_i\}, \tag{2.14}$$

where c_{i0} and c_{i1} are the weights for responses $X_i = 0$ and $X_i = 1$, respectively. Let

$$Z_i = c_{i0}(1 - X_i) + c_{i1}X_i, i = 1, 2, \ldots, I.$$

In this setup, we have

$$\mathrm{Var}(Z_i) = (c_{i0} - c_{i1})^2 \pi_i (1 - \pi_i), i = 1, 2, \ldots, I, \tag{2.15}$$

and

$$\mathrm{Cov}\left(Z_i, Z_j\right) = (c_{i0} - c_{i1})\left(c_{j0} - c_{j1}\right) \sum_{i=1}^{I} v_a (\pi_{ai} - \pi_i)\left(\pi_{aj} - \pi_j\right), i \neq j. \tag{2.16}$$

According to the above formulae, c_{i0} and c_{i1} are not identifiable, so we set $c_{i0} = 0$, $c_{i1} = c_i$, and $Z_i = c_i X_i, i = 1, 2, \ldots, I$. Then, the above formulae are rewritten as

$$\mathrm{Var}(Z_i) = {c_i}^2 \pi_i (1 - \pi_i) = {c_i}^2 \sum_{a=1}^{A} v_a \pi_{ai} (1 - \pi_{ai}) + {c_i}^2 \sum_{a=1}^{A} v_a (\pi_{ai} - \pi_i)^2, i = 1, 2, \ldots, I,$$

and

$$\mathrm{Cov}\left(Z_i, Z_j\right) = c_i c_j \sum_{a=1}^{A} v_a (\pi_{ai} - \pi_i)\left(\pi_{aj} - \pi_j\right), i \neq j.$$

Let

$$\sigma_{ijB} = \sum_{a=1}^{A} v_a(\pi_{ai} - \pi_i)(\pi_{aj} - \pi_j)$$

and

$$\sigma_{ijW} = \begin{cases} \sum_{a=1}^{A} v_a \pi_{ai}(1 - \pi_{ai}), i = j; \\ 0, i \neq j. \end{cases}$$

Then, the between- and within-class variance matrices of responses $\boldsymbol{X} = (X_1, X_2, \ldots, X_I)^T$ are defined by $\boldsymbol{\Sigma}_B = (\sigma_{ijB})$ and $\boldsymbol{\Sigma}_W = (\sigma_{ijW})$, respectively. In the above setup, the variance of score T (2.14) is calculated as follows:

$$\mathrm{Var}(T) = (c_1, c_2, \ldots, c_I)\boldsymbol{\Sigma}_B(c_1, c_2, \ldots, c_I)^T + (c_1, c_2, \ldots, c_I)\boldsymbol{\Sigma}_W(c_1, c_2, \ldots, c_I)^T. \tag{2.17}$$

The first term of the right-hand side of the above equation represents the between-class variance of $\boldsymbol{T}$ and the second term the within-class variance. Let

$$\mathrm{V}_B(T) = (c_1, c_2, \ldots, c_I)\boldsymbol{\Sigma}_B(c_1, c_2, \ldots, c_I)^T \tag{2.18}$$

and

$$\mathrm{V}_W(T) = (c_1, c_2, \ldots, c_I)\boldsymbol{\Sigma}_W(c_1, c_2, \ldots, c_I)^T. \tag{2.19}$$

For determining the weight vector $\boldsymbol{c} = (c_1, c_2, \ldots, c_I)$ that assesses the differences among the latent classes, the following criterion is used:

$$\frac{\mathrm{V}_B(T)}{\mathrm{V}_W(T)} \to \max_{\boldsymbol{c}} \frac{\mathrm{V}_B(T)}{\mathrm{V}_W(T)}. \tag{2.20}$$

In order to avoid the indeterminacy with respect to $(c_1, c_2, \ldots, c_I)$, we impose constraint

$$\mathrm{V}_W(T) = (c_1, c_2, \ldots, c_I)\boldsymbol{\Sigma}_W(c_1, c_2, \ldots, c_I)^T = 1 \tag{2.21}$$

on maximization (2.20). Then, the criterion is reduced to

$$\mathrm{V}_B(T) \to \max_{\boldsymbol{c}} \mathrm{V}_B(T). \tag{2.22}$$

Remark 2.2 In criterion (2.20), variance $\mathrm{V}_B(T)$ can be regarded as that according to latent classes. In this sense, it is interpreted as the signal variance of T, which is explained by A latent classes. On the other hand, the denominator $\mathrm{V}_W(T)$ can be viewed as the noise variance of T. Hence, the ratio

$$\frac{\mathrm{V}_B(T)}{\mathrm{V}_W(T)}$$

is the signal-to-noise ratio. In this sense, the above criterion is similar to KL information (1.29) and can be interpreted as entropy. ∎

In order to obtain the optimal weight vector $\boldsymbol{c} = (c_1, c_2, \ldots, c_I)^T$, the following Lagrange function is introduced:

$$g(\boldsymbol{c}) = (c_1, c_2, \ldots, c_I)\boldsymbol{\Sigma}_B(c_1, c_2, \ldots, c_I)^T - \lambda(c_1, c_2, \ldots, c_I)\boldsymbol{\Sigma}_W(c_1, c_2, \ldots, c_I)^T, \tag{2.23}$$

where λ is the Lagrange multiplier. Differentiating the above function with respect to vector $\boldsymbol{c} = (c_1, c_2, \ldots, c_I)$, we have

$$\frac{\partial g(\boldsymbol{c})}{\partial \boldsymbol{c}} = 2\boldsymbol{\Sigma}_B(c_1, c_2, \ldots, c_I)^T - 2\lambda\boldsymbol{\Sigma}_W(c_1, c_2, \ldots, c_I)^T = 0.$$

If $\boldsymbol{\Sigma}_W$ is non-singular, it follows that

$$\left(\boldsymbol{\Sigma}_W^{-1/2}\boldsymbol{\Sigma}_B\boldsymbol{\Sigma}_W^{-1/2} - \lambda\boldsymbol{E}\right)\boldsymbol{\Sigma}_W^{-1/2}(c_1, c_2, \ldots, c_I)^T = 0,$$

where $\boldsymbol{E}$ is the identity matrix of order I. Since $(c_1, c_2, \ldots, c_I) \neq 0$, λ is an eigenvalue of $\boldsymbol{\Sigma}_W^{-1/2}\boldsymbol{\Sigma}_B\boldsymbol{\Sigma}_W^{-1/2}$. Let K be the rank of $\boldsymbol{\Sigma}_W^{-1/2}\boldsymbol{\Sigma}_B\boldsymbol{\Sigma}_W^{-1/2}$; λ_k be the k th largest eigenvalues of the matrix; and let $\boldsymbol{\xi}_k$, $k = 1, 2, \ldots, K$ be the corresponding eigenvectors. Putting

$$\boldsymbol{c}_k = \boldsymbol{\Sigma}_W^{-1/2}\boldsymbol{\xi}_k,$$

$$T_k = (X_1, X_2, \ldots, X_I)\boldsymbol{c}_k, k = 1, 2, \ldots, K,$$

we have

$$\mathrm{V}_B(T_k) = \boldsymbol{c}_k\boldsymbol{\Sigma}_B\boldsymbol{c}_k^T = \boldsymbol{\xi}_k^T\boldsymbol{\Sigma}_W^{-1/2}\boldsymbol{\Sigma}_B\boldsymbol{\Sigma}_W^{-1/2}\boldsymbol{\xi}_k = \lambda_k, k = 1, 2, \ldots, K. \tag{2.24}$$

From (2.24), it is seen T_1 is the solution of (2.22). Since matrix $\boldsymbol{\Sigma}_W^{-1/2}\boldsymbol{\Sigma}_B\boldsymbol{\Sigma}_W^{-1/2}$ is symmetric, eigenvectors $\boldsymbol{\xi}_k$ are orthogonal with respect to the inner product, and thus, it follows that

$$\boldsymbol{c}_k \boldsymbol{\Sigma}_B \boldsymbol{c}_l^T = \boldsymbol{\xi}_k^T \boldsymbol{\Sigma}_W^{-1/2} \boldsymbol{\Sigma}_B \boldsymbol{\Sigma}_W^{-1/2} \boldsymbol{\xi}_l = 0, k \neq l.$$

From this, the weight vectors $\boldsymbol{c}_k, k = 1, 2, \ldots, K$ are orthogonal with respect to the between-variance matrix $\boldsymbol{\Sigma}_B$. The weight vectors make the scores or dimensions T_i to compare or order the latent classes. The locations of latent classes are based on dimensions T_i, that is,

$$t_{ak} \equiv E(T_k | \text{latent class } a) = (p_{a1}, p_{a2}, \ldots, p_{aI}) \boldsymbol{c}_k, k = 1, 2, \ldots, K; a = 1, 2, \ldots, A.$$

It is suitable to select two or three dimensions to express the locations of latent classes, that is, (t_{a1}, t_{a2}) or $(t_{a1}, t_{a2}, t_{a3}), a = 1, 2, \ldots, A$.

The above method for locating latent classes is demonstrated by the use of an artificial latent class model shown in Table 2.12. For this latent class model, the first- and second-best score functions (dimensions) T_1 and T_2 are derived according to eigenvalues (Table 2.13) and the locations of the latent classes are measured with the functions (Table 2.14). The score functions are interpreted according to the weights for manifest response variables $X_i, i = 1, 2, 3, 4$, and the locations of the latent classes are illustrated in Fig. 2.1. In the practical data analysis, it is an appropriate idea to interpret and compare the latent classes according to figures like Fig. 2.1.

According to locations (T_1, T_2) of latent classes in Table 2.14, the distances between latent classes are calculated. Let $d(a, b)$ be the Euclid distance between latent classes a and b. Then, we have

Table 2.12 A hypothesized latent class model

Latent class	Proportion	Positive response probability			
		X_1	X_2	X_3	X_4
1	0.2	0.1	0.5	0.3	0.7
2	0.5	0.8	0.6	0.1	0.2
3	0.3	0.9	0.3	0.7	0.3

Table 2.13 Score functions of the latent class model in Table 2.13

Score function	Eigenvalue	Weight			
		X_1	X_2	X_3	X_4
T_1	0.900	2.508	−0.151	0.468	−0.975
T_2	0.530	−0.245	−0.741	0.2302	0.615

Table 2.14 The two-dimensional scores of the three latent classes in Table 2.13

Latent Class	T_1	T_2
1	−0.366	0.726
2	1.768	−0.288
3	2.247	1.352

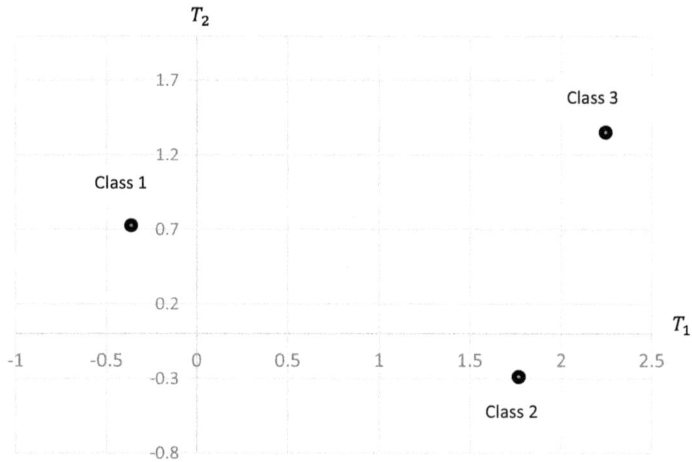

Fig. 2.1 Locations of latent classes in Table 2.13

$$d(1, 2) = 2.36, d(2, 3) = 1.71, d(1, 3) = 2.69. \tag{2.25}$$

The above Euclid distances between latent classes make a tree graph shown in Fig. 2.2.

Second, a method for comparing latent classes based on entropy is considered. For simplicity of the discussion, we discuss an entropy-based method for comparing latent classes in cases where the manifest variables X_i are binary. Let

$$\pi_{ai} = P(X_i = 1|a), i = 1, 2, \ldots, I; a = 1, 2, \ldots, A,$$

and let $\boldsymbol{p} = (p_1, p_2, \ldots, p_K)$ and $\boldsymbol{q} = (q_1, q_2, \ldots, q_K)$ be two probability distributions. Then, the divergences between the distributions are calculated as follows:

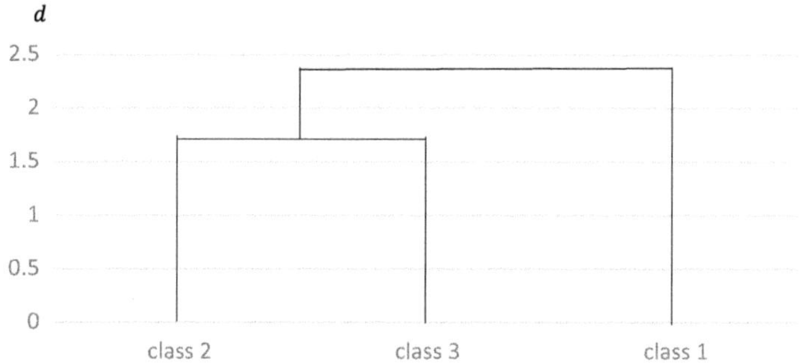

Fig. 2.2 The tree graph of latent classes in Table 2.12 based on the Euclidian distance

$$D(\boldsymbol{p}||\boldsymbol{q}) = \sum_{k=1}^{K} p_k \log\frac{p_k}{q_k},\ D(\boldsymbol{q}||\boldsymbol{p}) = \sum_{k=1}^{K} q_k \log\frac{q_k}{p_k}, \tag{2.26}$$

where

$$\sum_{k=1}^{K} p_k = \sum_{k=1}^{K} q_k = 1.$$

As in (1.29), the following KL information is used to measure the difference between the two distributions:

$$D^*(\boldsymbol{p}||\boldsymbol{q}) = D(\boldsymbol{p}||\boldsymbol{q}) + D(\boldsymbol{q}||\boldsymbol{p}). \tag{2.27}$$

From (2.26) and (2.27), we have

$$D^*(\boldsymbol{p}||\boldsymbol{q}) = \sum_{k=1}^{K} (p_k - q_k)(\log p_k - \log q_k)\mathrm{l}.$$

In model (2.1), the distribution of manifest variable vector $\boldsymbol{X} = (X_1, X_2, \ldots, X_I)^T$ in latent class a is

$$P(\boldsymbol{X} = \boldsymbol{x}|a) = \prod_{i=1}^{I} P(X_i = x_i|a). \tag{2.28}$$

Let $\boldsymbol{p}(\boldsymbol{X}|a)$ be the probability distribution of $\boldsymbol{X} = (X_1, X_2, \ldots, X_I)^T$ in latent class α and let $\boldsymbol{p}(X_i|a)$ be those of variables $X_i, i = 1, 2, \ldots, I$. Then, we have

$$D^*(\boldsymbol{p}(\boldsymbol{X}|a)||\boldsymbol{p}(\boldsymbol{X}|b)) = D(\boldsymbol{p}(\boldsymbol{X}|a)||\boldsymbol{p}(\boldsymbol{X}|b)) + D(\boldsymbol{p}(\boldsymbol{X}|b)||\boldsymbol{p}(\boldsymbol{X}|a)). \tag{2.29}$$

From (2.28), we also obtain

$$\begin{aligned} D^*(\boldsymbol{p}(\boldsymbol{X}|a)||\boldsymbol{p}(\boldsymbol{X}|b)) &= \sum_{i=1}^{I} D^*(\boldsymbol{p}(X_i|a)||\boldsymbol{p}(X_i|b)) \\ &= \sum_{i=1}^{I} \{D(\boldsymbol{p}(X_i|a)||\boldsymbol{p}(X_i|b)) + D(\boldsymbol{p}(X_i|b)||\boldsymbol{p}(X_i|a))\}. \end{aligned} \tag{2.30}$$

We see that the KL information concerning manifest variable vector $\boldsymbol{X}$ (2.29) is decomposed into I measures of KL information according to manifest variables $X_i, i = 1, 2, \ldots, I$. When manifest variables are binary, binary categories are, for

Table 2.15 KL distances between latent classes for variables X_i

	X_1	X_2	X_3	X_4
$D^*(\boldsymbol{p}(X_i\|1)\|\|\boldsymbol{p}(X_i\|2))$	2.51	0.04	0.27	1.12
$D^*(\boldsymbol{p}(X_i\|2)\|\|\boldsymbol{p}(X_i\|3))$	0.08	0.38	1.83	0.05
$D^*(\boldsymbol{p}(X_i\|1)\|\|\boldsymbol{p}(X_i\|3))$	3.52	0.17	0.68	0.68

example, {yes, no}, {positive, negative}, {success, failure}, and so on. Let π_{ai} be the positive responses of manifest variables $X_i, i = 1, 2, \ldots, I$ in latent classes $a = 1, 2, \ldots, A$. Then, (2.30) becomes

$$D^*(P(\boldsymbol{X}=\boldsymbol{x}|a)||P(\boldsymbol{X}=\boldsymbol{x}|b)) = \sum_{i=1}^{I}\left\{\begin{array}{l}\pi_{ai}\log\dfrac{\pi_{ai}}{\pi_{bi}} + (1-\pi_{ai})\log\dfrac{1-\pi_{ai}}{1-\pi_{bi}} + \pi_{bi}\log\dfrac{\pi_{bi}}{\pi_{ai}} \\ +(1-\pi_{bi})\log\dfrac{1-\pi_{bi}}{1-\pi_{ai}}\end{array}\right\}. \tag{2.31}$$

Applying the above results to Table 2.12, Table 2.15 illustrates the KL distances between latent classes for manifest variables X_i. From this table, we have

$$\begin{cases} D^*(P(\boldsymbol{X}=\boldsymbol{x}|1)||P(\boldsymbol{X}=\boldsymbol{x}|2)) = \sum_{i=1}^{4} D^*(\boldsymbol{p}(X_i|1)||\boldsymbol{p}(X_i|2)) = 3.94 \\ D^*(P(\boldsymbol{X}=\boldsymbol{x}|2)||P(\boldsymbol{X}=\boldsymbol{x}|3)) = 2.34, \\ D^*(P(\boldsymbol{X}=\boldsymbol{x}|1)||P(\boldsymbol{X}=\boldsymbol{x}|3)) = 5.04. \end{cases} \tag{2.32}$$

Based on the above measures, cluster analysis is used to compare the latent class model. Latent classes 2 and 3 are first combined, and the distance between {class 2, class 3} and class 1 is calculated by

$$\begin{aligned}&\min\{D^*(P(\boldsymbol{X}=\boldsymbol{x}|1)||P(\boldsymbol{X}=\boldsymbol{x}|2)), D^*(P(\boldsymbol{X}=\boldsymbol{x}|1)||P(\boldsymbol{X}=\boldsymbol{x}|3))\}\\ &\quad = D^*(P(\boldsymbol{X}=\boldsymbol{x}|1)||P(\boldsymbol{X}=\boldsymbol{x}|2)) = 3.94.\end{aligned}$$

From this, we have a tree graph shown in Fig. 2.3, and the result is similar to that in Fig. 2.2. As demonstrated above, the entropy-based method for comparing latent classes can be easily employed in data analyses.

Remark 2.3 From (2.31), we have

$$D^*(P(\boldsymbol{X}=\boldsymbol{x}|a)||P(\boldsymbol{X}=\boldsymbol{x}|b)) = \sum_{i=1}^{I}(\pi_{ai}-\pi_{bi})\left(\log\frac{\pi_{ai}}{1-\pi_{ai}} - \log\frac{\pi_{bi}}{1-\pi_{bi}}\right).$$

In a discussion similar to ECD in Sect. 1.6 (1.30), the above information (entropy) can be interpreted as a signal-to-noise ratio. In this case, the signal is $D^*(P(\boldsymbol{X}=\boldsymbol{x}|a)||P(\boldsymbol{X}=\boldsymbol{x}|b))$ and the noise is 1. From this, a standardized KL

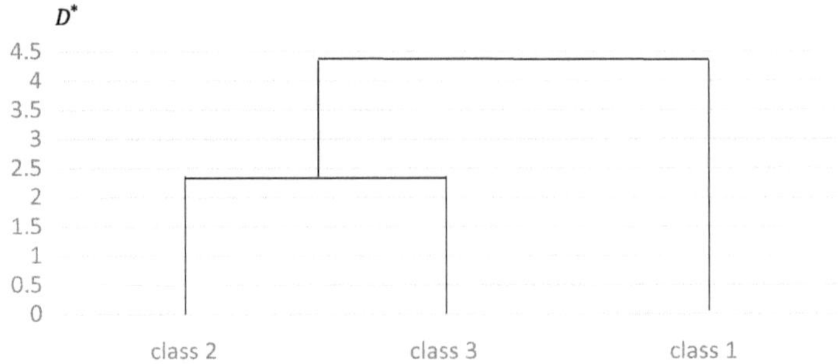

Fig. 2.3 The tree graph of latent classes in Table 2.12 based on KL information (2.32)

distance (2.31) can be defined by

$$\overline{D^*}(P(\boldsymbol{X} = \boldsymbol{x}|a)||P(\boldsymbol{X} = \boldsymbol{x}|b)) = \frac{D^*(P(\boldsymbol{X} = \boldsymbol{x}|a)||P(\boldsymbol{X} = \boldsymbol{x}|b))}{D^*(P(\boldsymbol{X} = \boldsymbol{x}|a)||P(\boldsymbol{X} = \boldsymbol{x}|b)) + 1}.$$

The interpretation of the above quantity can be given as that similar to ECD (1.30), i.e., the ratio of the variation of the two distributions in entropy.

2.6 Latent Profile Analysis

The latent profile model has been introduced in Sect. 1.4. In this section, an ML estimation procedure via the EM algorithm is constructed. Let $X_i, i = 1, 2, \ldots, I$ be manifest continuous variables; let X_{ai} be the latent variables in latent class $a = 1, 2, \ldots, A$, which are distributed according to normal distributions $N\left(\mu_{ai}, \psi_i^2\right)$; and let Z_a be the latent binary variables that take 1 for an individual in latent class a and 0 otherwise. Then, the model is expressed as

$$X_i = \sum_{a=1}^{A} Z_a X_{ai}, i = 1, 2, \ldots, I. \tag{2.33}$$

Let $x_{ij}, i = 1, 2, \ldots, I, j = 1, 2, \ldots, n$ be observed data that randomly selected individuals j take for manifest variables X_i, and let $Z_{aj} X_{aij}, i = 1, 2, \ldots, n, j = 1, 2, \ldots, n, a = 1, 2, \ldots, A$ be unobserved data that randomly selected individuals j in latent classes a take for latent variables $Z_a X_{ai}$.

Then, the incomplete and complete data are given, respectively, by

$$\begin{cases} Data = \left\{x_{ij}, i = 1, 2, \ldots, I, j = 1, 2, \ldots, n\right\}, \\ \mathfrak{D} = \left\{Z_{aj} X_{aij}, i = 1, 2, \ldots, n; a = 1, 2, \ldots, A\right\}, \end{cases} \tag{2.34}$$

where

$$x_{ij} = \sum_{a=1}^{A} Z_{aj} X_{aij}, i = 1, 2, \ldots, I;\ j = 1, 2, \ldots, n. \tag{2.35}$$

Let $\boldsymbol{\phi} = \left(v_a, \mu_{ai}, \psi_i^2\right)$ be the parameter vector, and ${}^s\boldsymbol{\phi} = \left({}^s v_a, {}^s\mu_{ai}, {}^s\psi_i^2\right)$ be the estimated parameters in the s th step in the EM algorithm. In order to construct the EM algorithm, the joint density function $f(\boldsymbol{x}|a)$ (1.16) is expressed by

$$f\left(\boldsymbol{x}|\left(\mu_{ai}, \psi_i^2\right)\right) \equiv f(\boldsymbol{x}|a) = \prod_{i=1}^{I} \frac{1}{\sqrt{2\pi\psi_i^2}} \exp\left\{-\frac{(x_i - \mu_{ai})^2}{2\psi_i^2}\right\}. \tag{2.36}$$

From (1.16), the log likelihood function of $\boldsymbol{\phi}$ based on the complete data $\mathfrak{D}$ is given by

$$\begin{aligned} \log l(\boldsymbol{\phi}|\mathfrak{D}) &= \sum_{a=1}^{A}\sum_{j=1}^{n} Z_{aj}\log v_a + n\sum_{i=1}^{I}\log\frac{1}{\sqrt{2\pi\psi_i^2}} - \sum_{a=1}^{A}\sum_{j=1}^{n}\sum_{i=1}^{I}\frac{Z_{aj}\left(x_{ij} - \mu_{ai}\right)^2}{2\psi_i^2} \\ &= \sum_{a=1}^{A}\sum_{j=1}^{n} Z_{aj}\log v_a - \frac{n}{2}\sum_{i=1}^{I}\log\psi_i^2 - \sum_{a=1}^{A}\sum_{j=1}^{n}\sum_{i=1}^{I}\frac{Z_{aj}\left(x_{ij} - \mu_{ai}\right)^2}{2\psi_i^2} - \frac{nI}{2}\log 2\pi. \end{aligned} \tag{2.37}$$

Let $\boldsymbol{x}_j = \left(x_{1j}, x_{2j}, \ldots, x_{Ij}\right), j = 1, 2, \ldots, n$ be the observed vectors for individuals j. Then, the EM algorithm is given as follows:

(i) E-step

For estimate ${}^s\boldsymbol{\phi}$ at the p th step, compute the conditional expectation of $\log f(\boldsymbol{x}, \boldsymbol{\xi}|\boldsymbol{\phi})$ given the incomplete data $\boldsymbol{X} = \boldsymbol{x}$ and parameter ${}^s\boldsymbol{\phi}$:

$$\begin{aligned} Q(\phi|{}^s\phi) &= \mathrm{E}\left(\log l(\phi|\mathfrak{D})|\boldsymbol{x}, {}^s\phi\right) = \sum_{a=1}^{A}\sum_{j=1}^{n} E\left(Z_{aj}|\boldsymbol{x}_j, {}^s\phi\right)\log v_a \\ &- \frac{n}{2}\sum_{i=1}^{I}\log\psi_i^2 - \sum_{a=1}^{A}\sum_{j=1}^{n}\sum_{i=1}^{I}\frac{E\left(Z_{aj}|\boldsymbol{x}_j, {}^s\phi\right)\left(x_{ij} - \mu_{ai}\right)^2}{2\psi_i^2} - \frac{nI}{2}\log 2\pi, \end{aligned} \tag{2.38}$$

where

$$E\left(Z_{aj}|\boldsymbol{x}_j, {}^s\boldsymbol{\phi}\right) = \frac{{}^s v_a f\left(\boldsymbol{x}_j|\left({}^s\mu_{ai}, {}^s\psi_i^2\right)\right)}{\sum_{a=1}^{A} {}^s v_a f\left(\boldsymbol{x}_j|\left({}^s\mu_{ai}, {}^s\psi_i^2\right)\right)}. \tag{2.39}$$

(ii) M-step

By using the Lagrange multiplier λ, the Lagrange function is given by

$$L(\boldsymbol{\phi}|{}^{s}\boldsymbol{\phi}) = Q(\boldsymbol{\phi}|{}^{s}\boldsymbol{\phi}) - \lambda \sum_{a=1}^{A} v_a. \tag{2.40}$$

Differentiating the above function with respect to v_a, we have

$$\frac{\partial}{\partial v_a} L(\boldsymbol{\phi}|{}^{s}\boldsymbol{\phi}) = \frac{1}{v_a} \sum_{j=1}^{n} E\left(Z_{aj}|\boldsymbol{x}, {}^{s}\boldsymbol{\phi}\right) - \lambda = 0, a = 1, 2, \ldots, A.$$

From the above equations, we obtain

$$\lambda = \sum_{a=1}^{A} \sum_{j=1}^{n} E\left(Z_{aj}|\boldsymbol{x}_j, {}^{s}\boldsymbol{\phi}\right) = n.$$

From this, it follows that

$${}^{s+1}v_a = \frac{1}{n} \sum_{j=1}^{n} E\left(Z_{aj}|\boldsymbol{x}_j, {}^{s}\boldsymbol{\phi}\right). \tag{2.41}$$

By differentiating (2.40), with respect to μ_{ai}, we get

$$\frac{\partial}{\partial \mu_{ai}} L(\boldsymbol{\phi}|{}^{s}\boldsymbol{\phi}) = \sum_{j=1}^{n} \frac{E\left(Z_{aj}|\boldsymbol{x}_j, {}^{s}\boldsymbol{\phi}\right)(x_{ij} - \mu_{ai})}{\psi_i^2} = 0, i = 1, 2, \ldots, I; a = 1, 2, \ldots, A.$$

From the above equations, we have

$${}^{s+1}\mu_{ai} = \frac{\sum_{j=1}^{n} x_{ij} E\left(Z_{aj}|\boldsymbol{x}_j, {}^{s}\boldsymbol{\phi}\right)}{\sum_{j=1}^{n} E\left(Z_{aj}|\boldsymbol{x}_j, {}^{s}\boldsymbol{\phi}\right)}, i = 1, 2, \ldots, I; a = 1, 2, \ldots, A. \tag{2.42}$$

Similarly, the partial differentiation of the Lagrange function with respect to ψ_i^2 gives

$$\frac{\partial}{\partial \psi_i^2} L(\boldsymbol{\phi}|{}^{s}\boldsymbol{\phi}) = -\frac{n}{2\psi_i^2} + \sum_{a=1}^{A} \sum_{j=1}^{n} \frac{E\left(Z_{aj}|\boldsymbol{x}_j, {}^{s}\boldsymbol{\phi}\right)\left(x_{ij} - {}^{s+1}\mu_{ai}\right)^2}{2\psi_i^4} = 0, i = 1, 2, \ldots, I.$$

From this, we have

$${}^{s+1}\psi_i^2 = \sum_{a=1}^{A} \sum_{j=1}^{n} \frac{E\left(Z_{aj}|\boldsymbol{x}_j, {}^{s}\boldsymbol{\phi}\right)\left(x_{ij} - {}^{s+1}\mu_{ai}\right)^2}{n}$$

$$= \frac{1}{n}\left(\sum_{j=1}^{n} x_{ij}^2 - \sum_{a=1}^{A} {}^{s+1}\mu_{ai}^2 \sum_{j=1}^{n} E\left(Z_{aj}|\boldsymbol{x}_j, {}^s\phi\right)\right), \ i = 1, 2, \ldots, I. \quad (2.43)$$

By using the above algorithm, the ML estimates of parameters in the latent profile model can be obtained. In some situations, we may relax the local independence of manifest variables, for example, correlations between some variables are assumed; however, overviewing the above process for constructing the EM algorithm, the modification is easy to make via a similar manner. In order to demonstrate the above algorithm, an artificial data set and the estimated parameters are given in Tables 2.16 and 2.17, respectively. The artificial data can be produced as a mixture of $\mathrm{N}(\boldsymbol{\mu}_1, \boldsymbol{\Psi})$ and $\mathrm{N}(\boldsymbol{\mu}_2, \boldsymbol{\Psi})$, where

$$\boldsymbol{\mu}_a = (\mu_{a1}, \mu_{a2}, \ldots, \mu_{a10})^T, a = 1, 2;$$

$$\boldsymbol{\Psi} = \begin{pmatrix} \psi_1^2 & 0 & \cdots 0 \\ 0 & \psi_2^2 & \cdots 0 \\ \vdots & \vdots & \vdots \ \vdots \\ 0 & 0 & \cdots \psi_{10}^2 \end{pmatrix}.$$

Remark 2.4 If in the latent profile model, the error variances of manifest variables X_i in latent classes a are different, that is, $\mathrm{Var}(X_i|a) = \psi_{ai}^2, i = 1, 2, \ldots, I$, then, the estimates in the EM algorithm are modified as follows:

$${}^{s+1}\psi_{ai}^2 = \frac{\sum_{j=1}^{n}\left(Z_{aj}|\boldsymbol{x}_j, {}^s\boldsymbol{\phi}\right)\left(x_{ij} - {}^{s+1}\mu_{ai}\right)^2}{\sum_{j=1}^{n} E\left(Z_{aj}|\boldsymbol{x}_j, {}^s\boldsymbol{\phi}\right)}, i = 1, 2, \ldots, I; a = 1, 2, \ldots, A.$$

In a framework of GLMs, the entropy coefficient of determination of the latent profile model can be calculated. The conditional density function of manifest variable vector $\boldsymbol{X}$, i.e., random component, is the following normal distribution:

$$f(\boldsymbol{x}|a) = \prod_{i=1}^{I} f_i(x_i|a) = \prod_{i=1}^{I} \frac{1}{\sqrt{2\pi\psi_i^2}} \exp\left\{-\frac{(x_i - \mu_i)^2}{2\psi_i^2}\right\},$$

where $f_i(x_i|a)$ are the conditional density functions in latent class a. As in the factor analysis model, we have

$$f_i(x_i|a) = \frac{1}{\sqrt{2\pi\psi_i^2}} \exp\left(\frac{x_i\mu_i - \frac{1}{2}\mu_i^2}{\psi_i^2} + \left(-\frac{x_i^2}{2\psi_i^2}\right)\right),$$

Table 2.16 Artificial data for a simulation

Number[a]	X_1	X_2	X_3	X_4	X_5	X_6	X_7	X_8	X_9	X_{10}
1	29.98	21.40	21.91	12.54	30.46	35.05	30.07	24.04	30.85	11.73
2	42.93	46.17	38.75	51.48	47.47	41.30	38.79	41.12	33.39	42.63
3	29.51	49.00	34.85	46.67	50.78	35.30	20.67	41.24	41.98	16.49
4	34.42	43.37	42.53	43.41	39.21	50.28	48.68	39.36	35.10	45.68
5	35.22	43.66	44.31	46.80	34.07	36.81	49.54	39.73	49.23	44.65
6	30.61	39.94	42.09	43.11	27.25	36.45	49.81	39.11	48.37	16.05
7	27.23	25.86	30.55	23.60	31.98	25.60	41.37	26.07	14.37	30.25
8	43.66	30.58	31.76	41.63	26.89	34.02	29.57	37.23	30.71	25.05
9	19.90	30.27	32.75	18.27	34.13	26.17	20.84	29.68	36.82	23.96
10	29.32	28.90	33.04	45.77	27.73	33.10	45.27	26.15	28.78	22.88
11	39.38	36.16	40.36	47.56	46.10	38.43	44.71	40.33	41.82	44.21
12	35.49	25.73	32.99	21.80	35.73	38.15	38.54	31.39	37.91	15.65
13	48.73	34.27	43.21	45.65	40.69	34.57	28.97	47.08	43.34	34.43
14	25.89	31.23	41.40	44.30	30.56	41.38	33.50	34.12	53.05	39.51
15	37.39	33.85	35.73	43.08	43.26	41.03	49.84	48.73	20.86	41.47
16	38.33	39.71	39.04	21.27	42.99	41.01	30.16	37.33	38.08	52.02
17	43.56	48.59	37.68	46.43	33.38	27.64	42.92	39.14	23.48	32.16
18	41.40	42.39	43.62	26.89	36.97	33.15	51.55	39.81	41.17	36.23
19	23.36	27.78	31.13	33.09	31.00	32.45	30.42	30.28	32.34	28.80
20	35.49	41.00	42.27	41.90	36.44	38.88	44.68	46.66	29.26	40.37
21	42.48	43.97	41.58	44.15	43.34	42.98	39.59	39.07	51.95	51.45
22	42.21	30.43	46.75	50.77	36.22	31.73	43.94	39.71	31.36	38.43
23	34.83	44.82	37.13	40.10	43.17	46.87	26.20	39.91	37.02	36.81
24	39.68	33.38	44.34	48.00	32.46	49.79	39.56	42.72	42.54	54.89
25	57.73	41.44	40.58	51.59	37.74	39.89	34.12	37.63	40.65	49.14
26	31.38	25.03	35.82	36.18	42.96	43.10	17.13	41.86	39.48	40.61
27	25.13	40.22	39.07	53.04	36.44	44.52	36.56	36.05	40.40	39.12
28	41.28	45.86	40.68	36.38	35.56	36.01	35.67	38.75	50.98	35.68
29	40.29	43.51	41.62	43.54	40.50	48.23	43.45	39.34	40.04	46.94
30	29.10	39.90	39.22	33.01	36.63	33.99	48.90	35.36	33.86	48.62
31	23.15	46.68	38.54	44.25	43.59	40.44	27.82	40.59	36.07	53.05
32	37.25	43.15	43.34	48.70	36.06	37.95	39.80	40.58	35.96	44.28
33	41.90	27.29	40.14	34.73	42.43	45.44	46.53	34.71	39.54	34.85
34	35.89	42.12	34.88	28.27	38.00	37.72	40.20	39.50	38.75	50.63
35	38.06	46.52	42.25	40.60	48.88	41.51	41.22	39.26	42.17	42.69
36	54.10	41.94	34.37	42.40	40.55	48.24	39.59	37.43	36.21	52.97

(continued)

Table 2.16 (continued)

Number[a]	X_1	X_2	X_3	X_4	X_5	X_6	X_7	X_8	X_9	X_{10}
37	34.51	39.71	36.92	34.76	45.32	41.96	40.77	37.48	29.30	45.20
38	43.49	29.91	40.68	35.27	37.08	44.86	45.41	35.67	41.09	52.50
39	33.28	48.02	39.74	49.58	46.76	35.46	42.19	46.45	31.18	45.67
40	43.19	41.79	40.63	34.80	40.94	33.25	45.27	41.52	38.99	44.07
41	41.84	38.39	42.78	42.70	29.65	28.38	47.16	40.80	36.83	39.81
42	39.58	35.49	40.89	41.51	44.39	40.39	36.03	41.40	39.28	24.35
43	45.69	30.53	39.37	46.24	32.83	42.10	40.59	37.02	39.02	40.29
44	32.33	51.63	41.19	43.05	37.27	56.24	26.20	46.26	39.84	26.44
45	35.71	46.13	44.59	28.89	34.21	42.95	35.78	37.53	33.84	44.41
46	48.10	41.86	41.00	34.68	43.57	47.68	37.73	39.94	30.89	32.79
47	36.24	42.08	40.92	31.25	46.97	32.84	47.80	42.12	35.10	51.52
48	41.80	38.64	38.81	36.58	37.25	37.30	26.64	41.30	44.83	41.18
49	29.41	26.24	43.38	32.65	45.29	34.81	39.09	38.78	43.61	59.67
50	36.18	39.57	41.21	35.43	34.36	33.46	33.51	44.07	43.94	53.88

[a] Number implies the data number produced for the simulation study

Table 2.17 The estimated parameters in a latent profile model with two latent classes

	X_1	X_2	X_3	X_4	X_5	X_6	X_7	X_8	X_9	X_{10}		
μ_{1i}	31.26	27.21	30.59	28.08	31.14	32.08	33.74	29.25	30.25	22.61	v_1	0.140
μ_{2i}	38.11	39.90	40.42	40.74	39.53	39.91	39.03	40.15	38.69	42.04	v_2	0.860
ψ_i^2	48.11	36.51	8.81	64.74	26.79	34.01	65.52	11.77	46.33	81.96	–	–

so we can set

$$\theta_i = \mu_i, a_i(\varphi_i) = \psi_i^2, b_i(\theta_i) = \frac{1}{2}\mu_i^2, c_i(x_i, \omega) = -\frac{x_i^2}{2\psi_i^2} - \log\sqrt{2\pi\psi_i^2}, i = 1, 2, \ldots, I.$$

Since the latent variable vector in the latent profile model is $\boldsymbol{Z} = (Z_1, Z_2, \ldots, Z_A)^T$, and the systematic components are given as follows:

$$\theta_i = \mu_i = \sum_{a=1}^{A} \mu_{ai} Z_a, i = 1, 2, \ldots, I.$$

Let $\boldsymbol{\theta} = (\theta_1, \theta_2, \ldots, \theta_I)^T$. Then, from (1.29) we have

$$\mathrm{KL}(\boldsymbol{X}, \boldsymbol{Z}) = \mathrm{tr}\Psi^{-1}\mathrm{Cov}(\boldsymbol{\theta}, \boldsymbol{X}) = \sum_{i=1}^{I} \frac{1}{\psi_i^2} \sum_{a=1}^{A} \mathrm{Cov}(\theta_i, X_i) = \sum_{i=1}^{I} \frac{1}{\psi_i^2} \sum_{a=1}^{A} \mathrm{Cov}(\mu_{ai} Z_a, X_i). \quad (2.44)$$

Since

$$E(X_i) = \sum_{a=1}^{A} v_a \mu_{ai},$$

we have

$$\sum_{a=1}^{A} \mathrm{Cov}(\mu_{ai} Z_a, X_i) = \sum_{a=1}^{A} v_a \mu_{ai}^2 - \overline{\mu}_i^2,$$

where

$$\overline{\mu}_i = \sum_{a=1}^{A} v_a \mu_{ai}, i = 1, 2, \ldots, I.$$

The entropy coefficient of determination (ECD) is calculated by

$$\mathrm{ECD}(\boldsymbol{X}, \boldsymbol{Z}) = \frac{\mathrm{KL}(\boldsymbol{X}, \boldsymbol{Z})}{\mathrm{KL}(\boldsymbol{X}, \boldsymbol{Z}) + 1}. \tag{2.45}$$

Similarly, we also have

$$\mathrm{ECD}(X_i, \boldsymbol{Z}) = \frac{\mathrm{KL}(X_i, \boldsymbol{Z})}{\mathrm{KL}(X_i, \boldsymbol{Z}) + 1}, \tag{2.46}$$

where

$$\mathrm{KL}(X_i, \boldsymbol{Z}) = \frac{1}{\psi_i^2} \sum_{a=1}^{A} \mathrm{Cov}(\mu_{ai} Z_a, X_i).$$

The information is that of $\boldsymbol{X}$ which the latent variable has. As in (2.9), the best way to classify observed data $\boldsymbol{X} = \boldsymbol{x}$ into the latent classes is based on the maximum posterior probability of $\boldsymbol{Z}$, that is, for

$$P(a_0 | (x_1, x_2, \ldots, x_I)) = \max_a \frac{v_a \prod_{i=1}^{I} f_i(x_i|a)}{\sum_{b=1}^{A} v_b \prod_{i=1}^{I} f_i(x_i|b)}, \tag{2.47}$$

an individual with response $\boldsymbol{x} = (x_1, x_2, \ldots, x_I)$ is evaluated as a member in latent class a_0.

The above discussion is applied to the estimated latent profile model shown in Table 2.17. Although in Table 2.18 there exist manifest variables that are less

Table 2.18 Assessment of the latent profile model with two latent classes in Table 2.18

	X_1	X_2	X_3	X_4	X_5	X_6	X_7	X_8	X_9	X_{10}	$\boldsymbol{X}$
KL	0.12	0.54	1.36	0.30	0.33	0.23	0.06	1.24	0.19	0.56	4.92
ECD	0.11	0.35	0.58	0.23	0.25	0.18	0.05	0.55	0.16	0.36	0.83

explained by the latent variable $\boldsymbol{Z}= (Z_1, Z_2)^T$, for example, X_1 and X_7, 83% of the variation of manifest variable vector $\boldsymbol{X}$ in entropy is explained by the latent variable. The ECDs in Table 2.18 are interpreted as the ratios of reduced uncertainty with respect to latent variable vector $\boldsymbol{Z}= (Z_1, Z_2)^T$, that is, latent classes. In effect, Table 2.19 shows the true and the estimated (assigned) latent classes of data in Table 2.16. Based on the true latent classes, the data in Table 2.16 have been made, and according to the estimated latent profile model, individuals are assigned to latent classes by using (2.47) (Table 2.19). The consistency ratio between true and estimated latent classes is $0.68(= 34/50)$.

Table 2.19 The true and assigned latent classes of individuals

Number[a]	1	2	3	4	5	6	7	8	9	10
LC[b]	2	2	2	2	2	2	2	2	2	1
ALC[c]	1	2	2	2	2	2	1	1	1	1
Number	11	12	13	14	15	16	17	18	19	20
LC	1	1	2	2	2	2	2	2	2	2
ALC	2	1	2	2	2	2	2	2	1	2
Number	21	22	23	24	25	26	27	28	29	30
LC	2	2	2	2	1	2	1	2	1	1
ALC	2	2	2	2	2	2	2	2	2	2
Number	31	32	33	34	35	36	37	38	39	40
LC	2	1	1	2	2	2	2	2	2	1
ALC	2	2	2	2	2	2	2	2	2	2
Number	41	42	43	44	45	46	47	48	49	50
LC	2	2	2	1	2	2	2	1	1	2
ALC	2	2	2	2	2	2	2	2	2	2

[a] Numbers imply those corresponding to data in Table 2.17
[b] LCs imply the true latent classes of the correspondent data in Table 2.17
[c] ALCs imply the latent classes of the correspondent data in Table 2.17, assigned with the estimated latent profile model in Table 2.18

2.7 Discussion

In this chapter, first, a general latent class analysis is discussed and for the ML estimation of the latent class model, the EM algorithm is constructed. For three data sets, latent class analysis has been demonstrated. Concerning the χ^2-test of the goodness-of-fit of the latent class model, the model fits the Stouffer-Toby and Lazarsfeld-Stouffer's data sets; however, we cannot have a good fit to McHugh's data set. Since the estimated latent class models for Stouffer-Toby's and Lazarsfeld-Stouffer's data sets have latent classes ordered as shown in Tables 2.2, 2.3, 2.8, and 2.9, it is meaningful to discuss latent class analysis assuming ordered latent classes, for example, the latent distance model. The basic latent class model treats latent classes parallelly, that is, without any assumption on latent response probabilities, and then, the analysis is called an exploratory latent class analysis or latent class cluster analysis [14, 21]. The number of latent classes in the latent class model is restricted by inequality (1.5), so to handle more ordered latent classes, it is needed to make parsimonious models as another approach. In order to assess the model performance, the explanatory power or goodness-of-fit of the model can be measured with ECD as demonstrated in Sect. 2.4. In the interpretation of latent classes, a method for locating the latent classes in a Euclidian space is given and the method is illustrated. An entropy-based method to compare the latent classes is also presented, and the method measures, in a sense, the KL distances between the latent classes, and the relationship among the latent classes is illustrated with cluster analysis. In Sect. 2.6, the latent profile model is considered, and the ML estimation procedure via the EM algorithm is constructed. A numerical illustration is given to demonstrate the latent profile analysis. These days, computer efficiency has been greatly increased, so the ML estimation procedures given in the present chapter can be realized in the EXCEL work files. The author recommends readers to make the calculations for the ML estimation of the latent class models for themselves. The present chapter has treated the basic latent class model, that is, an exploratory approach to latent class analysis. There may exist further studies to develop latent structure analysis, for example, making latent class model with ordered latent classes and extending the latent distance model (Lazarsfeld and Henry, 1968) and latent class models with explanatory variables [2]. Latent class analysis has also been applied in medical research [16, 17], besides in psychological and social science research [22]. In order to challenge confirmatory latent class approaches, it is important to extend research areas to apply the latent class model, and due to that, new latent structure models will be constructed to make effective and significant methods of latent structure analysis.

References

1. Anderson, T. W. (1954). On estimation of parameters in latent structure analysis. *Psychometrika, 19*, 1–10.
2. Dayton, C. M., & Macready, G. B. (1988). Concomitant-variable latent class models. *Journal of the American Statistical Association, 83*, 173–178.
3. Dempster, A. P., Laird, N. M., & Rubin, D. B. (1977). Maximum likelihood from incomplete data via the EM algorithm (with discussion). *Journal of the Royal Statistical Society, B, 39*, 1–38.
4. Eshima, N., & Tabata, M. (2010). Entropy coefficient of determination for generalized linear models. *Computational Statistics and Data Analysis, 54*, 1381–1389.
5. Forman, A. K. (1978). A note on parameter estimation for Lazarsfeld's latent class analysis. *Psychometrika, 43*, 123–126.
6. Gibson, W. A. (1955). An extension of Anderson's solution for the latent structure equations. *Psychometrika, 20*, 69–73.
7. Gibson, W. A. (1962). Extending latent class solutions to other variables. *Psychometrika, 27*, 73–81.
8. Goodman, L. A. (1974). Exploratory latent structure analysis using both identifiable and unidentifiable models. *Biometrika, 61*, 215–231.
9. Goodman, L. A. (1979). On the estimation of the parameters in latent structure analysis. *Psychometrika, 44*, 123–128.
10. Green, B. F. (1951). A general solution for the latent class model of latent structure analysis. *Psychometrika, 16*, 71–76.
11. Lazarsfeld, P. F. (1950). The logical and mathematical foundation of latent structure analysis. In Soufer, S. A., Guttman, L., & others (Eds.), *Measurement and prediction: Studies in social psychology I World War II* (Vol. 4). Prenceton University Press.
12. Lazarsfeld, P. F. & Henry, N. M. (1968). Latent Structure Analysis, Boston: Houghton Mifflin.
13. Madansky, A. (1960). Determinantal methods in latent class analysis. *Psychometrika, 25*, 183–198.
14. Magidson, J., & Vermunt, J. K. (2001). Latent class factor and cluster models: Bi-plots, and related graphical displays. *Sociological Methodology, 31*, 223–264.
15. McHugh, R. B. (1956). Efficient estimation of local identification in latent class analysis. *Psychometrika, 20*, 331–347.
16. Nosetti, L., Paglietti, M. G., Brunetti, L., Masini, L., Grutta, S. L., & Cilluffo, G. (2020). Application of latent class analysis in assessing the awareness, attitude, practice and satisfaction of paediatricians on sleep disorder management in children in Italy. *PLoS One, 15*(2), e0228377. https://doi.org/10.1371/journal.pone.0228377
17. Petersen, K. J., Qualter, P., & Humphery, N. (2019). The application of latent class analysis for investigating population child mental health: A systematic review. *Frontiers in Psychology, 10*, 1214. https://doi.org/10.3389/fpsyg.2019.01214.eCollection2019.
18. Price, L. C., Dayton, C. M., & Macready, G. B. (1980). Discovery algorithms for hierarchical relations. *Psychometrika, 45*, 449–465.
19. Proctor. C. H. (1970). A probabilistic formulation and statistical analysis of Guttman scaling, Psychometrika 35, 73-78.
20. Stouffer, S. A., & Toby, J. (1951). Role conflict and personality. *The American Journal of Sociology, 56*, 395–406.
21. Vermunt, J. K. (2010). Latent Class Models, *International Encyclopedia of. Education, 7*, 238–244.
22. Vermunt, J. K. (2003). Applications of latent class analysis in social science research. In Nielsen, T. D., Zhang, N. L. (Eds.), *Symbolic and quantitative approaches to reasoning with uncertainty. ECSQARU 2003*. Lecture Notes in Computer Science (Vol. 2711). Springer, Berlin, Heidelberg. https://doi.org/10.1007/978-3-540-45062-7_2.

Chapter 3
Latent Class Analysis with Ordered Latent Classes

3.1 Introduction

In latent class analysis with two latent classes, we can assume one class is higher than the other in a sense; however, for more than two latent classes, we cannot necessarily order them in one-dimensional sense. In such cases, to compare and interpret the latent classes, the dimensions of latent spaces to locate them are the number of latent classes minus one. In Chap. 2, a method for locating latent classes in a latent space has been considered, and the distances between latent classes are measured with the Euclidian distance in the latent space and then, cluster analysis is applied to compare the latent classes. Moreover, the Kullback–Leibler information (divergence) is also applied to measure the distances between the latent classes. Let $\pi_{ai}, i = 1, 2, \ldots, I; a = 1, 2, \ldots, A$ be the positive response probabilities to binary item X_i in latent class a. If

$$\pi_{1i} \le \pi_{2i} \le \cdots \le \pi_{Ai}, i = 1, 2, \ldots, I, \tag{3.1}$$

the latent classes can be ordered in one-dimensional concept or continuum. As shown in Examples in Sect. 2.3, latent class models with three latent classes have been estimated in Tables 2.2 and 2.8, and the estimated latent response probabilities are consistently ordered in the magnitudes as in (3.1), though the results came from exploratory latent class analyses, which are called latent class cluster analyses [11, 17]. In general, the explanatory latent class analysis cannot assure the consistency in order such as in (3.1) in parameter estimations. Two attempts for considering ordered latent classes were proposed in Lazarsfel and Henry [9], that is, the latent distance model that is an extension of the Guttman scaling analysis and a latent class model in which the latent classes are located in one-dimensional continuum by using polytomous functions for describing latent binary response probabilities π_{ai}. In the latter model, it is very difficult to estimate the model parameters with

N. Eshima, *An Introduction to Latent Class Analysis*, Behaviormetrics: Quantitative Approaches to Human Behavior 14,
https://doi.org/10.1007/978-981-19-0972-6_3

constraints $0 \leq \pi_{ai} \leq 1, i = 1, 2, \ldots, I; a = 1, 2, \ldots, A$. For latent distance analysis, the EM algorithm for the maximum likelihood estimation was given by Eshima and Asano [6] and by using a logit model instead of polytomous functions, ordered latent classes were treated by Eshima and Asano [7]. Croon [1] discussed latent class analysis with ordered latent classes for polytomous manifest variables by a non-parametric approach; however, the number of parameters is increasing as that of latent classes increasing, so in order to analyze ordered latent classes, it is better to use logit models for latent response probabilities, as shown in the subsequent sections. The approach can also be viewed as item response models with discrete latent traits. The ML estimation procedures for the Rasch models were discussed by De Leeuw and Verhelst [4] and Lindsay et al. [10].

In Sect. 3.2, latent distance model is discussed and an ML estimation procedure for the model parameters is constructed by the EM algorithm [6]. The procedure is demonstrated by using data sets used in Chap. 2. Section 3.3 discusses a method for assessing the latent Guttman scaling. In Sect. 3.4, the latent Guttman scaling is applied for discussing the association between two latent continuous traits. Section 3.5 provides an approach for dealing with ordered latent classes by the Rasch model [15]. In Sect. 3.6, a two-parameter latent trait model is treated and the ML estimation of the parameters is discussed through the EM algorithm [5]. Finally, Sect. 3.7 gives a discussion to lead to further studies.

3.2 Latent Distance Analysis

In the Guttman scaling, test items are ordered in the response difficulty and the purpose of the scaling is to evaluate the subjects under study in one-dimensional scale (trait or ability). Let $X_i, i = 1, 2, \ldots, I$ be responses to test items i, and let us set $X_i = 1$ for positive responses and $X_i = 0$ for negative ones. If $X_i = 1$, then, $X_{i-1} = 1, i = 2, 3, \ldots, I$ in the Guttman scaling, and thus, in a strict sense, there would be $I + 1$ response patterns, $(0, 0, \ldots, 0), (1, 0, \ldots, 0), \ldots, (1, 1, \ldots, 1)$, which we could observe; however, in the real observation, the other response patterns will also occur due to two kinds of response errors. One is the intrusion error and the other is the omission error. Hence, the Guttman scaling patterns can be regarded as skills with which the subjects can solve or respond successfully, for example, suppose that the ability of calculation in the arithmetic is measured by the following three items:

$$\text{(i) } X_1 : x + y =?, \quad \text{(ii) } X_2 : x \times y =?, \quad \text{and} \quad \text{(iii) } X_3 : x \div y =?,$$

then, the items are ordered in difficulty as the above order, and the response patterns to be observed would be $(0, 0, 0), (1, 0, 0), (1, 1, 0), (1, 1, 1)$. However, it is sensible to take the response errors into account. Let S_i be skills for manifest responses $X_i, i = 1, 2, \ldots, I$; let $S_i = 1$ be states of skill acquisitions for solving items i; $S_i = 0$ be those of non-skill acquisitions for items i and let us set.

$$P(X_i = 1|S_i = s_i) = \begin{cases} \pi_{Li} \ (s_i = 0) \\ \pi_{Hi} \ (s_i = 1) \end{cases}, i = 1, 2, \ldots, I. \tag{3.2}$$

Then, the intrusion error probabilities are π_{Li} and the omission error probabilities $1 - \pi_{Hi}$, and the following inequalities should hold:

$$0 < \pi_{Li} < \pi_{Hi} < 1, i = 1, 2, \ldots, I. \tag{3.3}$$

The latent classes and the positive response probabilities π_{Li} and π_{Hi} are illustrated in Table 3.1, where latent classes are denoted with the numbers of skill acquisitions, $\sum_{i=1}^{I} s_i$. In this sense, the latent classes are ordered in a hypothesized ability or trait, and it is significant for an individual with a response $\boldsymbol{x} = (x_1, x_2, \ldots, x_I)^T$ to assign to one of the latent classes, that is, an assessment of the individual's ability. In the latent distance model (3.2), the number of parameters is $3I$.

Remark 3.1 Term "skill" is used in the above explanation. Since it can be thought that skills represent thresholds in a continuous trait or ability to respond for test binary items, term "skill" is employed for convenience's sake in this book.

Overviewing the latent distance modes historically, the models are restricted versions of (2.2), considering the easiness of the parameter estimation. Proctor [14] proposed a model with

$$P(X_i = 1|S_i = s_i) = \begin{cases} \pi_L \quad (s_i = 0) \\ 1 - \pi_L \ (s_i = 1) \end{cases}, i = 1, 2, \ldots, I. \tag{3.4}$$

The intrusion and omission error probabilities are the same as π_L, constant through all items. Dayton and Macready [2] used the model with

$$P(X_i = 1|S_i = s_i) = \begin{cases} \pi_L \ (s_i = 0) \\ \pi_H \ (s_i = 1) \end{cases}, i = 1, 2, \ldots, I, \tag{3.5}$$

and the following improvement version of the above model was also proposed by Dayton and Macready [3]:

Table 3.1 Positive response probabilities in the latent distance model

Latent class	0	1	2	...	I
X_1	π_{L1}	π_{H1}	π_{H1}	...	π_{H1}
X_2	π_{L2}	π_{L2}	π_{H2}	...	π_{H2}
⋮	⋮	⋮	⋮	⋮	⋮
X_I	π_{LI}	π_{LI}	π_{LI}	...	π_{HI}

$$P(X_i = 1|S_i = s_i) = \begin{cases} \pi_{Li} & (s_i = 0) \\ 1 - \pi_{Li} & (s_i = 1) \end{cases}, i = 1, 2, \ldots, I. \tag{3.6}$$

In the above model, the intrusion error and omission error probabilities are the same as π_{Li} for items $i = 1, 2, \ldots, I$. The present model (3.2) with (3.3) is a general version of the above models.

In the present chapter, the ML estimation of model (3.2) with (3.3) is considered. For this model, the following reparameterization is employed [6]:

$$P(X_i = 1|S_i = s_i) = \begin{cases} \frac{\exp(\alpha_i)}{1+\exp(\alpha_i)}(= \pi_{Li}) & (s_i = 0) \\ \frac{\exp(\alpha_i+\exp(\beta_i))}{1+\exp(\alpha_i+\exp(\beta_i))}(= \pi_{Hi}) & (s_i = 1) \end{cases}, i = 1, 2, \ldots, I. \tag{3.7}$$

In this expression, the constraints (3.3) are satisfied. The above model expression can be simplified as follows:

$$P(X_i = 1|S_i = s_i) = \frac{\exp(\alpha_i + s_i\exp(\beta_i))}{1 + \exp(\alpha_i + s_i\exp(\beta_i))}, i = 1, 2, \ldots, I. \tag{3.8}$$

Let $\boldsymbol{S} = (S_1, S_2, .., S_I)^T$ be a latent response (skill acquisition) vector and let $\boldsymbol{X} = (X_1, X_2, .., X_I)^T$ a manifest response vector. Then, the latent classes corresponding to latent response $\boldsymbol{s} = (s_1, s_2, .., s_I)^T$ can be described with score $k = \sum_{i=1}^{I} s_i$. From (2.1), we have

$$\begin{aligned} &P(\boldsymbol{X} = \boldsymbol{x}|\boldsymbol{S} = \boldsymbol{s}) \\ &= \prod_{i=1}^{I}\left(\frac{\exp(\alpha_i + s_i\exp(\beta_i))}{1 + \exp(\alpha_i + s_i\exp(\beta_i))}\right)^{x_i}\left(\frac{1}{1 + \exp(\alpha_i + s_i\exp(\beta_i))}\right)^{1-x_i} \\ &= \prod_{i=1}^{I}\frac{\exp\{x_i(\alpha_i + s_i\exp(\beta_i))\}}{1 + \exp(\alpha_i + s_i\exp(\beta_i))} \end{aligned}$$

and

$$P(\boldsymbol{X} = \boldsymbol{x}) = \sum_{k=0}^{I} v_k P(\boldsymbol{X} = \boldsymbol{x}|k) = \sum_{k=0}^{I} v_k \prod_{i=1}^{I}\frac{\exp\{x_i(\alpha_i + s_i\exp(\beta_i))\}}{1 + \exp(\alpha_i + s_i\exp(\beta_i))}.$$

In order to estimate the parameters $\boldsymbol{\phi} = ((v_k), (\alpha_i), (\beta_i))^T$, the following EM algorithm is used.

EM algorithm I

(i) E-step

Let ${}^s\boldsymbol{\phi} = (({}^s v_k), ({}^s\alpha_i), ({}^s\beta_i))^T$ be the estimate of parameter vector $\boldsymbol{\phi}$ at the s th iteration in the EM algorithm. Then, in the $(s+1)$ th iteration, the conditional expectations of complete data $(n(\boldsymbol{x}, k))$ for given parameters ${}^s\boldsymbol{\phi} = (({}^s v_k), ({}^s\alpha_i), ({}^s\beta_i))$ are calculated as follows:

$$ {}^{s+1}n(\boldsymbol{x}, k) = n(\boldsymbol{x}) \frac{{}^s v_k \prod_{i=1}^{I} {}^s P(X_i = x_i|k)}{\sum_{m=0}^{I} {}^s v_m \prod_{i=1}^{I} {}^s P(X_i = x_i|m)}, k = 0, 1, 2, \ldots, I, \tag{3.9} $$

where

$$ {}^s P(X_i = x_i|k) = \frac{\exp\{x_i({}^s\alpha_i + s_i \exp({}^s\beta_i))\}}{1 + \exp({}^s\alpha_i + s_i \exp({}^s\beta_i))}, x_i = 0, 1. $$

(ii) M-step

By using the complete data (3.9), the loglikelihood function based on the complete data $\left({}^{s+1}n(\boldsymbol{x}, k)\right)$ is given by

$$ \begin{aligned} Q(\boldsymbol{\phi}|{}^s\boldsymbol{\phi}) &= l\left(\boldsymbol{\phi}|\left({}^{s+1}n(\boldsymbol{x}, k)\right)\right) \\ &= \sum_{k=0}^{I} \sum_{\boldsymbol{x}} {}^{s+1}n(\boldsymbol{x}, k) \log\left(v_k \prod_{i=1}^{I} \frac{\exp\{x_i(\alpha_i + s_i \exp(\beta_i))\}}{1 + \exp(\alpha_i + s_i \exp(\beta_i))} \right) \end{aligned} $$

$$ \begin{aligned} &= \sum_{k=0}^{I} \sum_{\boldsymbol{x}} {}^{s+1}n(\boldsymbol{x}, k) \log v_k \\ &+ \sum_{k=0}^{I} \sum_{\boldsymbol{x}} {}^{s+1}n(\boldsymbol{x}, k) \left[\sum_{i=1}^{I} \{x_i(\alpha_i + s_i \exp(\beta_i)) - \log(1 + \exp(\alpha_i + s_i \exp(\beta_i)))\} \right]. \end{aligned} \tag{3.10} $$

With respect to ${}^{s+1}v_k$, as in (2.10), we have

$$ {}^{s+1}v_k = \frac{\sum_{\boldsymbol{x}} {}^{s+1}n(\boldsymbol{x}, k)}{\lambda} = \frac{\sum_{\boldsymbol{x}} {}^{s+1}n(\boldsymbol{x}, k)}{N}, k = 0, 1, 2, \ldots, I; $$

however, the other parameters α_i and β_i cannot be obtained explicitly, so we have to use the Newton–Raphson method for maximizing $Q(\boldsymbol{\phi}|{}^s\boldsymbol{\phi})$ in the M-step. The first derivatives of $Q(\boldsymbol{\phi}|{}^s\boldsymbol{\phi})$ with respect to α_i and β_i, respectively, are calculated as follows:

$$ \frac{\partial Q(\boldsymbol{\phi}|{}^s\boldsymbol{\phi})}{\partial \alpha_i} = \sum_{k=0}^{I} \sum_{\boldsymbol{x}} {}^{s+1}n(\boldsymbol{x}, k) \left(x_i - \frac{\exp(\alpha_i + s_i \exp(\beta_i))}{1 + \exp(\alpha_i + s_i \exp(\beta_i))} \right) $$

$$= \sum_{k=0}^{I} \sum_{\boldsymbol{x}} {}^{s+1}n(\boldsymbol{x}, k)(x_i - P(X_i = 1|S_i = s_i)), i = 1, 2, \ldots, I;$$

$$\frac{\partial Q(\boldsymbol{\phi}|^s\boldsymbol{\phi})}{\partial \beta_i} = \sum_{k=0}^{I} \sum_{\boldsymbol{x}} {}^{s+1}n(\boldsymbol{x}, k)(x_i - P(X_i = 1|S_i = s_i))s_i\exp(\beta_i), i = 1, 2, \ldots, I.$$

Then, the $2I$-dimensional gradient vector is set as

$$\boldsymbol{g} = \left(\begin{pmatrix} \frac{\partial Q(\boldsymbol{\phi}|^s\boldsymbol{\phi})}{\partial \alpha_i} \end{pmatrix} \atop \begin{pmatrix} \frac{\partial Q(\boldsymbol{\phi}|^s\boldsymbol{\phi})}{\partial \alpha_i} \end{pmatrix} \right). \tag{3.11}$$

Consequently, the second-order partial derivatives of $Q(\boldsymbol{\phi}|^s\boldsymbol{\phi})$ are calculated as follows:

$$\frac{\partial^2 Q(\boldsymbol{\phi}|^s\boldsymbol{\phi})}{\partial \alpha_i^2} = -\sum_{k=0}^{I} \sum_{\boldsymbol{x}} {}^{s+1}n(\boldsymbol{x}, k)P(X_i = 1|S_i = s_i)(1 - P(X_i = 1|S_i = s_i)),$$

$$i = 1, 2, \ldots, I;$$

$$\frac{\partial^2 Q(\boldsymbol{\phi}|^s\boldsymbol{\phi})}{\partial \alpha_i \partial \beta_i} = -\sum_{k=0}^{I} \sum_{\boldsymbol{x}} {}^{s+1}n(\boldsymbol{x}, k)P(X_i = 1|S_i = s_i)(1 - P(X_i = 1|S_i = s_i))s_i\exp(\beta_i),$$

$$i = 1, 2, \ldots, I;$$

$$\frac{\partial^2 Q(\boldsymbol{\phi}|^s\boldsymbol{\phi})}{\partial \beta_i^2} = \sum_{k=0}^{I} \sum_{\boldsymbol{x}} {}^{s+1}n(\boldsymbol{x}, k)$$

$$\{x_i - P(X_i = 1|S_i = s_i) - P(X_i = 1|S_i = s_i)(1 - P(X_i = 1|S_i = s_i))s_i\exp(\beta_i)\}s_i\exp(\beta_i),$$

$$i = 1, 2, \ldots, I;$$

$$\frac{\partial^2 Q(\boldsymbol{\phi}|^s\boldsymbol{\phi})}{\partial \alpha_i \partial \alpha_j} = \frac{\partial^2 Q(\boldsymbol{\phi}|^s\boldsymbol{\phi})}{\partial \alpha_i \partial \beta_j} = \frac{\partial^2 Q(\boldsymbol{\phi}|^s\boldsymbol{\phi})}{\partial \beta_i \partial \beta_j} = 0, i \neq j.$$

From the above results, the Hessian matrix $\boldsymbol{H}$ is set as follows:

$$\boldsymbol{H} = \begin{pmatrix} \left(\frac{\partial^2 Q(\boldsymbol{\phi}|^s\boldsymbol{\phi})}{\partial \alpha_i \partial \alpha_j}\right) & \left(\frac{\partial^2 Q(\boldsymbol{\phi}|^s\boldsymbol{\phi})}{\partial \alpha_i \partial \beta_i}\right) \\ \left(\frac{\partial^2 Q(\boldsymbol{\phi}|^s\boldsymbol{\phi})}{\partial \beta_i \partial \alpha_i}\right) & \left(\frac{\partial^2 Q(\boldsymbol{\phi}|^s\boldsymbol{\phi})}{\partial \beta_i \partial \beta_j}\right) \end{pmatrix}. \tag{3.12}$$

Let $\boldsymbol{\phi}_{(\alpha,\beta)} = ((\alpha_i), (\beta_i))$ and let t th iterative value of $\boldsymbol{\phi}_{(\alpha,\beta)}$ be $\boldsymbol{\phi}_{(\alpha,\beta)t} = ((\alpha_{it}), (\beta_{it}))$, where $\boldsymbol{\phi}_{(\alpha,\beta)1} = ((^s\alpha_i), (^s\beta_i))$, Then, $\boldsymbol{\phi}_{(\alpha,\beta)t+1}$ is obtained as follows:

$$\boldsymbol{\phi}_{(\alpha,\beta)t+1} = \boldsymbol{\phi}_{(\alpha,\beta)t} - \boldsymbol{H}_t^{-1}\boldsymbol{g}_t, t = 1, 2, \ldots, \tag{3.13}$$

where $\boldsymbol{H}_t$ and $\boldsymbol{g}_t$ are values of the gradient vector (3.11) and the Hessian matrix (3.12) at $\boldsymbol{\phi} = \boldsymbol{\phi}_t = \left(\left({}^{s+1}v_k\right), \boldsymbol{\phi}_{(\alpha,\beta)t}\right)$, respectively. From this algorithm, we can get $\lim_{t\to\infty}\boldsymbol{\phi}_{(\alpha,\beta)t} = \left(\left({}^{s+1}\alpha_i\right), \left({}^{s+1}\beta_i\right)\right)$.

Remark 3.2 The Newton–Raphson method for obtaining the estimates ${}^{s+1}\alpha_i$ and ${}^{s+1}\beta_i$ in the M-step makes a quick convergence of sequence $\boldsymbol{\phi}_{(\alpha,\beta)t}$ within several iterations.

Remark 3.3 Without constraints in (3.3), the latent distance model is a latent class model with the following equality constraints:

$$\begin{cases} P(X_i = 1|0) = P(X_i = 1|1) = \cdots P(X_i = 1|i-1)(= \pi_{Li}), \\ P(X_i = 1|i) = P(X_i = 1|i+1) = \cdots P(X_i = 1|I)(= \pi_{Hi}), \end{cases} i = 1, 2, \ldots, I.$$

Then, the EM algorithm can be applied for estimating the parameters. Let $\boldsymbol{\phi} = ((v_a), (\pi_{Li}), (\pi_{Hi}))^T$ be the parameters to be estimated and let ${}^s\boldsymbol{\phi} = (({}^sv_k), ({}^s\pi_{Li}), ({}^s\pi_{Hi}))^T$ be the estimates of the parameters at the s th iteration. Then, the EM algorithm is given as follows:

EM algorithm II

(i) E-step

$${}^{s+1}n(\boldsymbol{x}, k) = n(\boldsymbol{x})\frac{{}^sv_k\prod_{i=1}^I {}^sP(X_i = x_i|a)}{\sum_{m=0}^I {}^sv_m\prod_{i=1}^I {}^sP(X_i = x_i|b)}, k = 0, 1, 2, \ldots, I, \tag{3.14}$$

where from (3.7)

$${}^sP(X_i = 1|k) = \begin{cases} {}^s\pi_{Li} & (a < i) \\ {}^s\pi_{Hi} & (a \geq i) \end{cases}, k = 0, 1, 2, \ldots, I.$$

(ii) M-step

$${}^{s+1}v_k = \frac{\sum_{\boldsymbol{x}} {}^{s+1}n(\boldsymbol{x}, k)}{\lambda} = \frac{\sum_{\boldsymbol{x}} {}^{s+1}n(\boldsymbol{x}, k)}{N}, k = 0, 1, 2, \ldots, I;$$

$${}^{s+1}\hat{\pi}_{Hi} = \frac{1}{N}\sum_{k=i}^{I}\sum_{\boldsymbol{x}} {}^{s+1}n(\boldsymbol{x}, k)x_i, i = 1, 2, \ldots, I; \tag{3.15}$$

$${}^{s+1}\hat{\pi}_{Li} = \frac{1}{N}\sum_{k=0}^{i-1}\sum_{\boldsymbol{x}} {}^{s+1}n(\boldsymbol{x}, k)x_i, i = 1, 2, \ldots, I. \tag{3.16}$$

The algorithm is a proportional fitting one; however, the results do not necessarily guarantee the inequality constraints in (3.3). ■

The data in Table 2.1 is analyzed by using the latent distance model. The data are from respondents to questionnaire items on role conflict [16], and the positive responses $X_i, i = 1, 2, 3, 4$ are 171, 108, 111, and 67, respectively. It may be valid that item 1 is the easiest and item 4 is the most difficult to obtain positive responses, whereas items 2 and 3 are intermediate. The estimated class proportions and the positive response probabilities are given in Table 3.2. From a test of the goodness-of-fit to the data is very good. The assessment of responses in the five latent classes is illustrated in Table 3.3, and the results are compared with the response scores $\sum_1^4 x_i$. Assuming a latent continuum in a population, the estimated item response probabilities in the latent distance model and the five latent classes are illustrated in Fig. 3.1. As demonstrated in this example, it is significant to grade the respondents with response patterns instead of simple scores, for example, response patterns (1, 1, 1, 0), (1, 1, 0, 1), (1, 0, 1, 1), and (0, 1, 1, 1) have manifest score 3; however, they are assigned to latent classes 3, 4, 1, 0, respectively.

Table 3.2 Results of latent distance analysis of the data in Table 2.1

Latent class	Proportion	Latent positive item response probability			
		X_1	X_2	X_3	X_4
0	0.296	0.324	0.253	0.364	0.136
1	0.344	0.988	0.253	0.364	0.136
2	0.103	0.988	0.940	0.364	0.136
3	0.049	0.988	0.940	0.948	0.136
4	0.208	0.988	0.940	0.948	0.973

$G^2 = 0.921(df = 3, P = 0.845)$

Table 3.3 Assignment of the manifest responses to the extracted latent classes (latent distance analysis of data set in Table 2.1)

Response pattern	Score[a]	Latent class	Response pattern	Score	Latent class
0 0 0 0	0	0	0 0 0 1	1	0
1 0 0 0	1	1	1 0 0 1	2	1
0 1 0 0	1	0	0 1 0 1	2	0
1 1 0 0	2	2	1 1 0 1	3	4
0 0 1 0	1	0	0 0 1 1	2	0
1 0 1 0	2	1	1 0 1 1	3	1
0 1 1 0	2	0	0 1 1 1	3	0
1 1 1 0	3	3	1 1 1 1	4	4

[a]Scores imply the sums of the positive responses

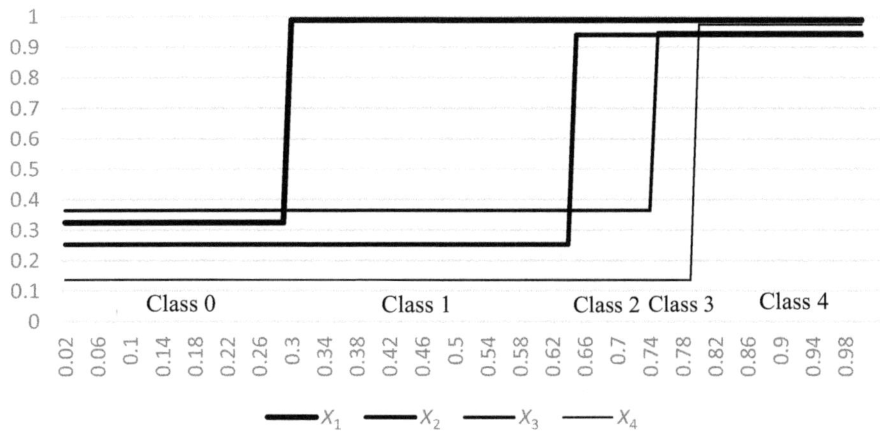

Fig. 3.1 Graph of the estimated latent distance model for data set in Table 2.1

In order to assess the goodness-of-fit of the latent distance model, that is, the explanatory power, the entropy approach with (2.12) and (2.13) is applied to the above analysis. Comparing Tables 2.10, 2.11, and 3.4, the goodness-of-fit of the latent distance model is better than the other models. From Table 3.4, 70% of the entropy of response variable vector $\boldsymbol{X} = (X_1, X_2, X_3, X_4)^T$ are explained by the five ordered latent classes in the latent distance model, and item 1 (X_1) is associated with the latent variable stronger than the other manifest variables.

Data in Tables 2.5 (McHugh's data) and those in Table 2.7 (Lazarsfel-Stouffer's data) are also analyzed with the latent distance model. The first data were obtained from four test items on creative ability in machine design [12], and the second data were from noncommissioned officers that were cross-classified with respect to their dichotomous responses, "favorable" and "unfavorable" toward the Army for each of the four different items on general attitude toward the Army [13]. Before analyzing the data sets, the marginal frequencies of positive response to items are given in Table 3.5. Considering the marginal positive response frequencies, in McHugh's data set it is natural to think there are no orders in difficulty with respect to item responses X_i; whereas in Lazarsfel-Stouffer's data set (Table 2.7) it may be appropriate to assume the difficulty order in the item responses, i.e., the skill acquisition order

$$S_1 \prec S_2 \prec S_3 \prec S_4.$$

Table 3.4 Assessment of the latent distance model for the Stouffer-Toby data

Manifest variable	X_1	X_2	X_3	X_4	Total
KL	0.718	0.606	0.386	0.625	2.335
ECD	0.418	0.377	0.278	0.385	0.700

Table 3.5 Marginal positive response frequencies of McHugh's and Lazarsfel-Stouffer's data

Data set	Marginal positive response frequency			
	X_1	X_2	X_3	X_4
McHugh's data	65	75	78	73
Lazarsfeld-Stouffer's data	359	626	700	736

The results of latent distance analysis of Lazarsfel-Stouffer's data set are given in Table 3.6 and the estimated model is illustrated in Fig. 3.2. The goodness-of-fit of the model to the data set is not statistically significant at the level of significance 0.05, and comparing the results with those in Table 2.8 or Table 2.9, the latter is better to explain the data set. Figure 3.2 demonstrates the estimated latent distance model, and the entropy-based assessment of the latent distance model is illustrated in Table 3.7.

The Guttman scaling is an efficient method to grade subjects with their response patterns; however, in the practical observation or experiments, we have to take their

Table 3.6 Results of latent distance analysis of the data set in Table 2.7

Latent class	Proportion	Latent positive item response probability			
		X_1	X_2	X_3	X_4
0	0.388	0.027	0.366	0.445	0.498
1	0.030	0.569	0.366	0.445	0.498
2	0.038	0.569	0.813	0.445	0.498
3	0.031	0.569	0.813	0.914	0.498
4	0.513	0.569	0.813	0.914	0.981

$G^2 = 6.298(df = 3,\ P = 0.098)$

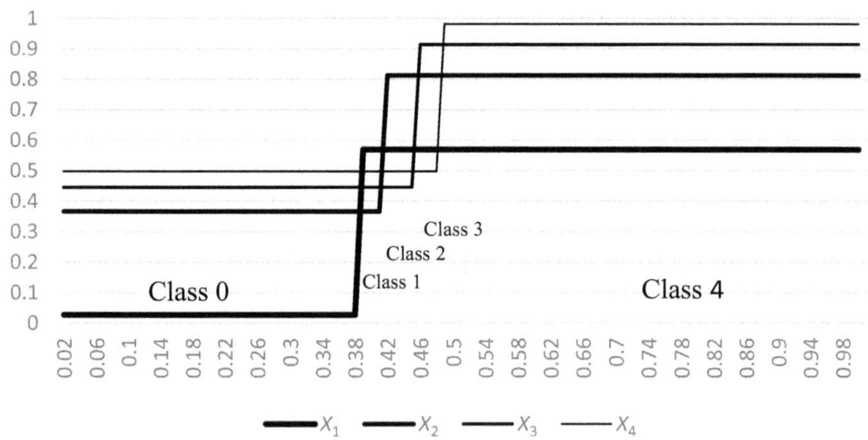

Fig. 3.2 Graph of the estimated latent distance model for data set in Table 2.7

Table 3.7 Assessment of the latent distance model for Lazarsfel-Stouffer's Data

Manifest variable	X_1	X_2	X_3	X_4	Total
KL	0.496	0.219	0.300	0.478	1.493
ECD	0.332	0.179	0.231	0.324	0.599

response errors into account. In this respect, the latent distance model provides a good approach to deal with the method. The approach is referred to as the latent Guttman scaling in this book. In applying latent distance models to data sets, the contents of items to be used have to be considered beforehand.

Remark 3.4 Setting initial estimates of π_{Li} and π_{Hi} satisfying the constraints in (3.3), if the estimates in the latent distance model by EM algorithm II satisfy the same constraints, they are the same as those by EM algorithm I.

3.3 Assessment of the Latent Guttman Scaling

Let $X_1, X_2, \ldots, X_I$ be manifest variables that make the latent Guttman scaling. As in the previous discussion, the manifest variables are observed to assess the latent continuous trait θ that is distributed according to the uniform distribution on interval $[0, 1]$. In the latent distance model, we have $I + 1$ ordered latent classes, and it is significant to assess the latent classes with scores on the interval, that is, locate them on the interval. The assessment of the trait by using the latent Guttman scale depends on the items to be employed, that is, the latent distance model, and so it is meaningful to discuss the information about trait θ that the model has [8]. The amount of information implies the model performance, that is, goodness of the scaling. Let

$$\theta_{(0)} = 0, \theta_{(i)} = \sum_{k=0}^{i-1} v_k, i = 1, 2, \ldots, I + 1. \tag{3.17}$$

Then, $\theta_{(i)}$ are interpreted as the threshold for positively or successfully responding to item i, that is,$X_i = 1, i = 1, 2, \ldots, I$. Let us assign the following scores to latent classes i:

$$T(\theta) = \begin{cases} t_i & \left(\theta \in \left[\theta_{(i)}, \theta_{(i+1)}\right), i = 0, 1, \ldots, I - 1\right) \\ t_I & \left(\theta \in \left[\theta_{(I)}, 1\right]\right) \end{cases}. \tag{3.18}$$

The information ratio about latent trait θ, which score $T(\theta)$ has, is defined by

$$\mathcal{K}(T(\theta)|\theta) \equiv \text{corr}(T(\theta), \theta)^2 = \frac{\text{Cov}(T(\theta), \theta)^2}{\text{Var}(T(\theta))\text{Var}(\theta)}. \tag{3.19}$$

From the above definition, we have

$$0 < \mathcal{K}(T(\theta)|\theta) < 1.$$

Since θ is uniformly distributed on interval [0, 1], we have

$$\mathrm{E}(\theta) = \frac{1}{2}, \ \mathrm{var}(\theta) = \frac{1}{12}.$$

From (3.18), we also get

$$\mathrm{E}(T(\theta)) = \sum_{k=0}^{I} t_k v_k = \sum_{k=0}^{I} t_k \left(\theta_{(k+1)} - \theta_{(k)}\right), \tag{3.20}$$

$$\mathrm{Var}(T(\theta)) = \sum_{k=0}^{I} t_k^2 \left(\theta_{(k+1)} - \theta_{(k)}\right) - \mathrm{E}(T(\theta))^2, \tag{3.21}$$

$$\mathrm{Cov}(T(\theta), \theta) = \frac{1}{2}\left\{\sum_{k=0}^{I} t_k \left(\theta_{(k+1)}^2 - \theta_{(k)}^2\right) - \mathrm{E}(T(\theta))\right\}. \tag{3.22}$$

By using the above results, we obtain

$$\mathcal{K}(T(\theta)|\theta) = \frac{12\mathrm{Cov}(T(\theta), \theta)^2}{\mathrm{Var}(T(\theta))} = \frac{3\left\{\sum_{k=0}^{I} t_k \left(\theta_{(k+1)}^2 - \theta_{(k)}^2\right) - \mathrm{E}(T(\theta))\right\}^2}{\mathrm{Var}(T(\theta))}. \tag{3.23}$$

The amount of information about latent trait θ that the manifest variables have is defined by

$$\mathcal{K}(X_1, X_2, \ldots, X_I|\theta) = \max_{T(\theta)\in\mathcal{F}} \mathcal{K}(T(\theta)|\theta), \tag{3.24}$$

where $\mathcal{F}$ is the class of functions defined by (3.18). We have the following theorem:

Theorem 3.1 *Let* θ *be uniformly distributed on interval* [0, 1], *and let function* $T(\theta)$ *be defined by* (3.18). *Then,* $\mathcal{K}(T(\theta)|\theta)$ *is maximized by*

$$t_i = a\frac{\theta_{i+1} + \theta_i}{2} + b, i = 0, 1, 2, \ldots, I, \tag{3.25}$$

where a and b are constant, and it follows that

$$\mathcal{K}(X_1, X_2, \ldots, X_I|\theta) = 3\sum_{i=0}^{I} \theta_{i+1}\theta_i(\theta_{i+1} - \theta_i). \tag{3.26}$$

Proof In order to maximize (3.23) with respect to $T(\theta)$, the following constraints are imposed on the function, that is, normalization:

$$\mathrm{E}(T(\theta)) = \sum_{a=0}^{I} t_i \left(\theta_{(i+1)} - \theta_{(i)}\right) = 0, \tag{3.27}$$

$$\mathrm{Var}(T(\theta)) = \sum_{i=0}^{I} t_i^2 \left(\theta_{(i+1)} - \theta_{(i)}\right) - \mathrm{E}(T(\theta))^2 = 1. \tag{3.28}$$

From (3.27), we have

$$\mathrm{Var}(T(\theta)) = \sum_{i=0}^{I} t_i^2 \left(\theta_{(i+1)} - \theta_{(i)}\right) = 1. \tag{3.29}$$

From constraints (3.27) and (3.28), it follows that

$$\mathcal{K}(T(\theta)|\theta) = 3\left\{\sum_{i=0}^{I} t_i \left(\theta_{(i+1)}^2 - \theta_{(i)}^2\right)\right\}^2.$$

In order to make the maximization of the above function with respect to scores t_i, it is sufficient to maximize the following one:

$$\sum_{i=0}^{I} t_i \left(\theta_{(i+1)}^2 - \theta_{(i)}^2\right).$$

For Lagrange multipliers λ and μ, the following Lagrange function is made:

$$g = \sum_{i=0}^{I} t_i \left(\theta_{(i+1)}^2 - \theta_{(i)}^2\right) - \lambda \sum_{i=0}^{I} t_i \left(\theta_{(i+1)} - \theta_{(i)}\right) - \mu \sum_{a=0}^{I} t_i^2 \left(\theta_{(i+1)} - \theta_{(i)}\right).$$

Differentiating the above function with respect to t_i, we have

$$g = \sum_{i=0}^{I} t_i \left(\theta_{(i+1)}^2 - \theta_{(i)}^2\right) - \lambda \sum_{i=0}^{I} t_i \left(\theta_{(i+1)} - \theta_{(i)}\right) - \mu \sum_{a=0}^{I} t_i^2 \left(\theta_{(i+1)} - \theta_{(i)}\right). \tag{3.30}$$

From this,

$$\left(\theta_{(i+1)} - \theta_{(i)}\right)\left(\theta_{(i+1)} + \theta_{(i)} - \lambda - 2\mu t_i\right) = 0.$$

Since $\theta_{(i+1)} - \theta_{(i)} \neq 0$, we have

$$\theta_{(i+1)} + \theta_{(i)} - \lambda - 2\mu t_i = 0. \tag{3.31}$$

Summing up both sides of (3.30) with respect to $i = 0, 1, 2, \ldots, I$, it follows that

$$1 - \lambda - 2\mu \sum_{i=0}^{I} t_i \left(\theta_{(i+1)} - \theta_{(i)}\right) = 0. \tag{3.32}$$

From (3.27), we have

$$\lambda = 1,$$

and from (3.31) we get

$$t_i = \frac{\theta_{(i+1)} + \theta_{(i)} - 1}{2\mu}, i = 0, 1, 2, \ldots, I. \tag{3.33}$$

Multiplying (3.30) by t_i and summing up both sides of the equations with respect to $i = 0, 1, 2, \ldots, I$, we have

$$\sum_{i=0}^{I} t_i \left(\theta_{(i+1)}^2 - \theta_{(i)}^2\right) - \lambda \sum_{i=0}^{I} t_i \left(\theta_{(i+1)} - \theta_{(i)}\right) - 2\mu \sum_{i=0}^{I} t_i^2 \left(\theta_{(i+1)} - \theta_{(i)}\right) = 0.$$

From (3.27) and (3.29), it follows that

$$\sum_{i=0}^{I} t_i \left(\theta_{(i+1)}^2 - \theta_{(i)}^2\right) - 2\mu = 0.$$

From the above equation, we have

$$\mu = \frac{\sum_{i=0}^{I} t_i \left(\theta_{(i+1)}^2 - \theta_{(i)}^2\right)}{2}. \tag{3.34}$$

From (3.33) and (3.34), we get

$$\mu = \frac{\sum_{i=0}^{I} \left(\theta_{(i+1)} + \theta_{(i)} - 1\right)\left(\theta_{(i+1)}^2 - \theta_{(i)}^2\right)}{4\mu} = \frac{\sum_{i=0}^{I} \theta_{(i+1)}\theta_{(i)}\left(\theta_{(i+1)} - \theta_{(i)}\right)}{4\mu}.$$

By solving the above equation with respect to $\mu(> 0)$, we have

$$\mu = \frac{\sqrt{\sum_{i=0}^{I} \theta_{(i+1)}\theta_{(i)}\left(\theta_{(i+1)} - \theta_{(i)}\right)}}{2}$$

and (3.23) is maximized by $T(\theta)$ with (3.33), that is,

$$\begin{aligned}\mathcal{K}(X_1, X_2, \ldots, X_I|\theta) &= \max_{T(\theta)\in\mathcal{F}} \mathcal{K}(T(\theta)|\theta)\\ &= 3\left\{\frac{\sum_{i=0}^{I}\left(\theta_{(i+1)} + \theta_{(i)} - 1\right)\left(\theta_{(i+1)}^2 - \theta_{(i)}^2\right)}{2\mu}\right\}^2\\ &= 3\sum_{i=0}^{I}\theta_{(i+1)}\theta_{(i)}\left(\theta_{(i+1)} - \theta_{(i)}\right). \qquad (3.35)\end{aligned}$$

Since $\mathcal{K}(T(\theta)|\theta)$ (3.19) is the square of the correlation coefficient between $T(\theta)$ and θ, hence the theorem follows. ■

From Theorem 3.1, we set

$$t_i = \frac{\theta_{(i+1)} + \theta_{(i)}}{2}, i = 0, 1, 2, \ldots, I. \qquad (3.36)$$

The above discussion is applied to the latent distance models estimated in Tables 3.2 and 3.6. For Table 3.2, we have

$$\theta_{(0)} = 0, \theta_{(1)} = 0.296, \theta_{(2)} = 0.640, \theta_{(3)} = 0.743, \theta_{(4)} = 0.792, \theta_{(4)} = 1,$$

and from (3.33) it follows that

$$\mathcal{K}(X_1, X_2, X_3, X_4|\theta) = 0.923.$$

Similarly, for Table 3.6, we obtain

$$\mathcal{K}(X_1, X_2, X_3, X_4|\theta) = 0.806.$$

From the results, the latent Guttman scaling in Table 3.2 is better than that in Table 3.6.

The following theorem gives the maximization of (3.24) with respect to $\theta_{(i)}, i = 0, 1, 2, \ldots, I$.

Theorem 3.2 *The amount of information about latent trait* θ, $\mathcal{K}(X_1, X_2, \ldots, X_I|\theta)$, *is maximized with respect to* $\theta_{(i)}, i = 0, 1, 2, \ldots, I$ *by*

$$\theta_{(i)} = \frac{i}{I+1}, i = 0, 1, 2, \ldots, I$$

and then, it follows that

$$\max_{(\theta_{(a)})} \mathcal{K}(X_1, X_2, \ldots, X_I|\theta) = \frac{I(I+2)}{(I+1)^2}. \tag{3.37}$$

Proof Differentiating $\mathcal{K}(X_1, X_2, \ldots, X_I|\theta)$ with respect to $\theta_{(i)}$, we have

$$\frac{\partial}{\partial\theta_{(i)}}\mathcal{K}(X_1, X_2, \ldots, X_I|\theta) = 3\big(\theta_{(i+1)} - \theta_{(i-1)}\big)\big(\theta_{(i+1)} + \theta_{(i-1)} - 2\theta_{(i)}\big) = 0.$$

Since $\theta_{(i+1)} \neq \theta_{(i-1)}$, we obtain

$$\theta_{(i+1)} + \theta_{(i-1)} - 2\theta_{(i)} = 0.$$

Therefore, it follows that

$$\theta_{(i)} = \frac{i}{I+1}, i = 0, 1, 2, \ldots, I+1. \tag{3.38}$$

From this, we get (3.35) and the theorem follows. ■

By using the above theorem, the efficiency of the latent Guttman scaling can be defined by

$$efficiency = \frac{\mathcal{K}(X_1, X_2, \ldots, X_I|\theta)}{\max_{(\theta_{(a)})} \mathcal{K}(X_1, X_2, \ldots, X_I|\theta)}. \tag{3.39}$$

The efficiencies of the latent distance models in Tables 3.2 and 3.6 are calculated, respectively, as 0.962 and 0.840.

Remark 3.5 The efficiency of the latent Guttman scaling may also be measured with entropy. Let $\boldsymbol{p} = (p_1, , p_2, , \ldots, p_A)$ be any probability distribution. Then, the entropy is defined by

$$H(\boldsymbol{p}) = -\sum_{a=1}^{A} p_a \log p_a.$$

The maximum of the above entropy is $\log A$ for the uniform distribution $\boldsymbol{q} = \left(\frac{1}{A}, \frac{1}{A}, \ldots, \frac{1}{A}\right)$. The result is the same as that in Theorem 3.2. Then, the efficiency of distribution $\boldsymbol{p}$ can be defined by

$$efficiency = \frac{H(\boldsymbol{p})}{\log A}.$$

Applying the above efficiency to the latent distance models estimated in Tables 3.2 and 3.6, we have 0.892 and 0.650, respectively. In the sense of entropy, the latent Guttman scaling in Table 3.2 is better than that in Table 3.6 as illustrated above by using (3.39).

3.4 Analysis of the Association Between Two Latent Traits with Latent Guttman Scaling

The latent distance model discussed in Sect. 3.2 is extended to a multidimensional version to measure the association between latent traits [8]. Let X_{ki} be binary items for measuring the acquisition of skills $S_{ki}, i = 1, 2, \ldots, I_k, k = 1, 2$ for hierarchically assessing continuous latent traits $\theta_k, k = 1, 2$, which are ordered as $S_{k1} \prec S_{k2} \prec \ldots S_{kI_k}$ by the difficulty of the skill acquisition in trait $\theta_k, k = 1, 2$. For simplicity of the notations, let us set $\boldsymbol{X}_k = \left(X_{k1}, X_{k2}, \ldots, X_{kI_k}\right)$ and $\boldsymbol{S}_k = \left(S_{k1}, S_{k2}, \ldots, S_{kI_k}\right), k = 1, 2$. In this setting, as in the previous section,

$$X_{ki} = \begin{cases} 1 \ (success) \\ 0 \ (failure) \end{cases}, S_{ki} = \begin{cases} 1 \quad\quad (acquition) \\ 0 \ (nonacquisition) \end{cases}, i = 1, 2, \ldots, I_k; k = 1, 2.$$

In this setup, the skills $S_{k1} \prec S_{k2} \prec \ldots S_{kI_k}$ constitute the latent Guttman scaling. Let $\theta_{k(a)}$ be thresholds for skills $S_{ka}, a = 0, 1, 2, \ldots, I_k + 1; k = 1, 2$, and then, we set

$$v_{mn} = P\left(\theta_{1(m)} \le \theta_1 < \theta_{1(m+1)}, \theta_{2(n)} \le \theta_2 < \theta_{2(n+1)}\right),$$

$$m = 0, 1, 2, \ldots, I_1; n = 0, 1, 2, \ldots, I_2.$$

Then, putting $\boldsymbol{s}_k = \left(s_{k1}, s_{k2}, \ldots, s_{kI_k}\right), k = 1, 2$, and

$$m = \sum_{i=0}^{I_1} s_{1i}, n = \sum_{i=0}^{I_2} s_{2i},$$

the model is given by

$$P((\boldsymbol{X}_1, \boldsymbol{X}_2) = (\boldsymbol{x}_1, \boldsymbol{x}_2)) = \sum_{k=0}^{I_1} \sum_{l=0}^{I_2} v_{mn} P((\boldsymbol{X}_1, \boldsymbol{X}_2) = (\boldsymbol{x}_1, \boldsymbol{x}_2) | (\boldsymbol{S}_1, \boldsymbol{S}_2) = (\boldsymbol{s}_1, \boldsymbol{s}_2)),$$

where

$$\begin{aligned} &P((\boldsymbol{X}_1, \boldsymbol{X}_2) = (\boldsymbol{x}_1, \boldsymbol{x}_2) | (\boldsymbol{S}_1, \boldsymbol{S}_2) = (\boldsymbol{s}_1, \boldsymbol{s}_2)) \\ &= \prod_{k=1}^{2} \prod_{i=1}^{I_k} \frac{\exp\{x_{ki}(\alpha_{ki} + s_{ki}\exp(\beta_{ki}))\}}{1 + \exp(\alpha_{ki} + s_{ki}\exp(\beta_{ki}))}. \end{aligned}$$

According to the model, the joint levels of traits θ_k of individuals can be scaled, and the association between the traits can also be assessed. Let $T_k(\theta_k), k = 1, 2$ be functions of scores for latent traits θ_k, which are made by (3.18) and (3.36). Then, the correlation coefficient between scores $T_k(\theta_k), k = 1, 2$, $\mathrm{Corr}(T_1(\theta_1), T_2(\theta_2))$ is used for measuring the association between traits θ_1 and θ_2, because $\mathrm{Corr}(\theta_1, \theta_2)$ cannot be calculated. If θ_1 and θ_2 are statistically independent, $T_1(\theta_1)$ and $T_2(\theta_2)$ are also independent, and then, we have $\mathrm{Corr}(T_1(\theta_1), T_2(\theta_2)) = 0$.

The above model is applied to data in Table 3.8, which were obtained from 145 children from 1 to 5 years old. Latent trait θ_1 and θ_2 implied the general intelligence and the verbal ability of children, respectively, and these abilities are measured with three manifest binary variables ordered as $X_{ki}, i = 1, 2, 3; k = 1, 2$, respectively. The parameter can be estimated via the EM algorithm as in the previous section. The estimated latent probabilities are illustrated in Table 3.9, and the responses to the manifest variables X_{ki} are demonstrated in Figs. 3.3 and 3.4. From Fig. 3.3, we have

$$\mathcal{K}(X_1, X_2, X_3|\theta_1) = 0.840, \ efficiency = 0.896.$$

Similarly, for Table Fig. 3.4, it follows that

$$\mathcal{K}(X_1, X_2, X_3|\theta_2) = 0.898, \ efficiency = 0.958.$$

In this data, the mean densities of domains $\left[\theta_{1(m)}, \theta_{1(m+1)}\right) \times [\theta_{2(n)}.\theta_{2(n+1)})$ are calculated as

$$\frac{v_{mn}}{\left(\theta_{1(m+1)} - \theta_{1(m)}\right) \times \left(\theta_{2(n+1)} - \theta_{2(n)}\right)}, m, n = 0, 1, 23.$$

The densities are illustrated in Fig. 3.5, and the association between traits.θ_1 and θ_2 is summarized. It seems the association is positive, and in effect we obtain estimate $\widehat{\mathrm{Corr}}(T_1(\theta_1), T_2(\theta_2)) = 0.780$. From this, the association between the two latent traits is strong. The respondents shown in Table 3.8 are assigned to latent classes in Table 3.10, and it implies an assessment of respondents' grades in the latent traits.

In this section, two-dimensional latent continuous traits are discretized, and ordering of latent classes can be carried out in each latent trait; however, it may be useful to grade all the latent classes with a method, because without it, for latent classes $(i.j), i = 0, 1, 2, 3; j = 0, 1, 2, 3$ we may simply employ scores $i + j$ to grade the latent classes. In Sect. 4.10 in Chap. 4, an entropy-based method to order latent classes is discussed, and grading (ordering) of the above latent classes $(i.j), i = 0, 1, 2, 3; j = 0, 1, 2, 3$ (Table 3.9) will be treated as an example.

3.5 Latent Ordered-Class Analysis

In analyzing Stouffer-Toby's data (Table 2.1), latent class cluster analysis and latent distance analysis have been used. From the results in Tables 2.3 and 2.11, it is appropriate to assume there exist ordered latent classes that explain behavior in the

Table 3.8 Data on the general intelligence ability and the verbal ability from 145 pupils

θ_1			θ_2				θ_1			θ_2				θ_1			θ_2			
X_{11}	X_{12}	X_{13}	X_{21}	X_{22}	X_{23}	Freq	X_{11}	X_{12}	X_{13}	X_{21}	X_{22}	X_{23}	Freq	X_{11}	X_{12}	X_{13}	X_{21}	X_{22}	X_{23}	Freq
0	0	0	0	0	0	13	0	1	1	0	1	0	0	0	0	1	1	0	1	0
1	0	0	0	0	0	5	1	1	1	0	1	0	1	1	0	1	1	0	1	2
0	1	0	0	0	0	1	0	0	0	1	1	0	1	0	1	1	1	0	1	0
1	1	0	0	0	0	2	1	0	0	1	1	0	7	1	1	1	1	0	1	5
0	0	1	0	0	0	1	0	1	0	1	1	0	3	0	0	0	0	1	1	0
1	0	1	0	0	0	2	1	1	0	1	1	0	5	1	0	0	0	1	1	0
0	1	1	0	0	0	0	0	0	1	1	1	0	0	0	1	0	0	1	1	0
1	1	1	0	0	0	2	1	0	1	1	1	0	2	1	1	0	0	1	1	1
0	0	0	1	0	0	9	0	1	1	1	1	0	2	0	0	1	0	1	1	0
1	0	0	1	0	0	3	1	1	1	1	1	0	11	1	0	1	0	1	1	0
0	1	0	1	0	0	0	0	0	0	0	0	1	0	0	1	1	0	1	1	1
1	1	0	1	0	0	2	1	0	0	0	0	1	0	1	1	1	0	1	1	1
0	0	1	1	0	0	0	0	1	0	0	0	1	0	0	0	0	1	1	1	0
1	0	1	1	0	0	2	1	1	0	0	0	1	0	1	0	0	1	1	1	1
0	1	1	1	0	0	1	0	0	1	0	0	1	1	0	1	0	1	1	1	0

(continued)

Table 3.8 (continued)

θ_1			θ_2				θ_1			θ_2				θ_1			θ_2			
X_{11}	X_{12}	X_{13}	X_{21}	X_{22}	X_{23}	Freq	X_{11}	X_{12}	X_{13}	X_{21}	X_{22}	X_{23}	Freq	X_{11}	X_{12}	X_{13}	X_{21}	X_{22}	X_{23}	Freq
1	1	1	1	0	0	4	1	0	1	0	0	1	0	1	1	0	1	1	1	3
0	0	0	0	1	0	1	0	1	1	0	0	1	0	0	0	1	1	1	1	0
1	0	0	0	1	0	1	1	1	1	0	0	1	0	1	0	1	1	1	1	2
0	1	0	0	1	0	1	0	0	0	1	0	1	0	0	1	1	1	1	1	3
1	1	0	0	1	0	1	1	0	0	1	0	1	1	1	1	1	1	1	1	38
0	0	1	0	1	0	0	0	1	0	1	0	1	0							
1	0	1	0	1	0	2	1	1	0	1	0	1	1							

Source Eshima [8]

Table 3.9 The estimated latent probabilities (parameters)

Latent class (m, n)	Class proportion	Positive response probabilities to items					
		X_{11}	X_{12}	X_{13}	X_{21}	X_{22}	X_{23}
(0,0)	0.134	0.120	0.060	0.154	0.073	0.113	0.031
(1,0)	0.051	0.894	0.060	0.154	0.073	0.113	0.031
(2,0)	0.020	0.864	0.914	0.154	0.073	0.113	0.031
(3,0)	0.013	0.894	0.914	0.947	0.073	0.113	0.031
(0,1)	0.068	0.120	0.060	0.154	0.931	0.113	0.031
(1,1)	0.020	0.894	0.060	0.154	0.931	0.113	0.031
(2,1)	0.002	0.894	0.914	0.154	0.931	0.113	0.031
(3,1)	0.028	0.894	0.914	0.947	0.931	0.113	0.031
(0,2)	0.000	0.120	0.060	0.154	0.931	0.856	0.031
(1,2)	0.071	0.894	0.060	0.154	0.931	0.856	0.031
(2,2)	0.080	0.894	0.914	0.154	0.931	0.856	0.031
(3,2)	0.090	0.894	0.914	0.947	0.931	0.856	0.031
(0,3)	0.000	0.120	0.060	0.154	0.931	0.856	0.936
(1,3)	0.008	0.894	0.060	0.154	0.931	0.856	0.936
(2,3)	0.019	0.894	0.914	0.154	0.931	0.856	0.936
(3,3)	0.396	0.894	0.914	0.947	0.931	0.856	0.936

loglikelohood ratio statistic $= 39.507$, $df = 39$, $p = 0.447$

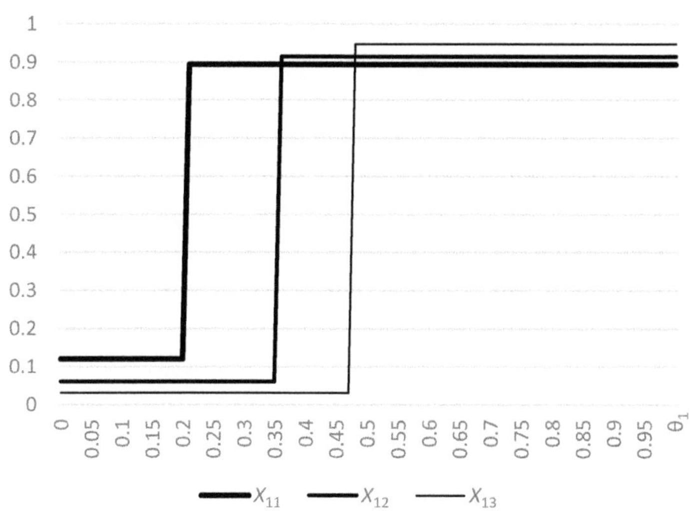

Fig. 3.3 The estimated response probabilities to X_{11}, X_{12}, and X_{13} for measuring θ_1

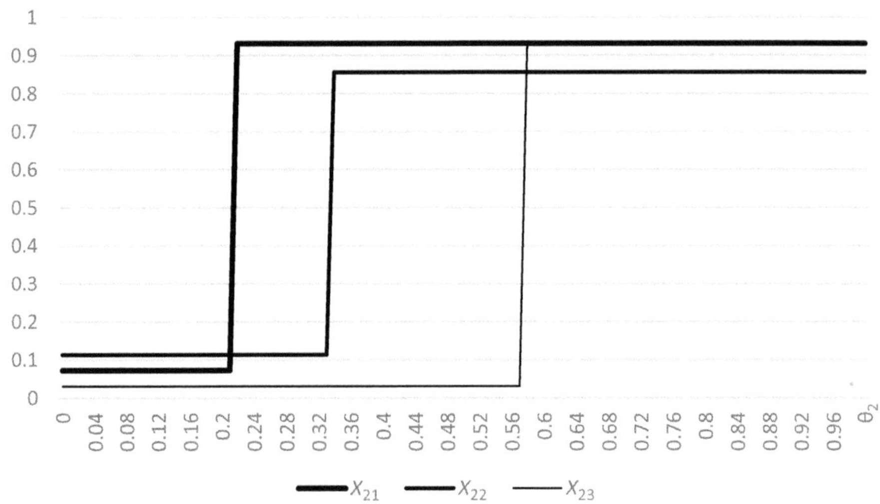

Fig. 3.4 The estimated response probabilities to X_{21}, X_{22}, and X_{23} for measuring θ_2

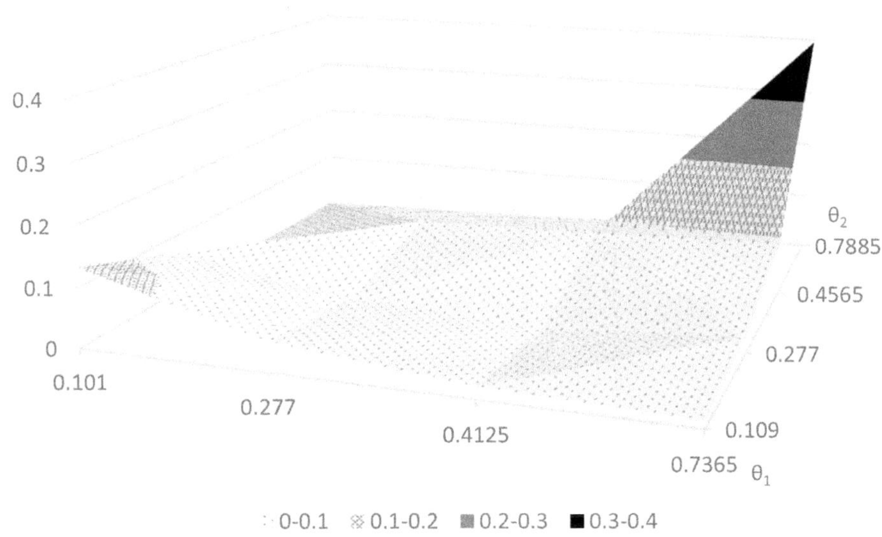

Fig. 3.5 Summaries of mean densities between traits θ_1 and θ_2

data set, that is, role conflict. The results in Table 2.3 are those based on a latent class cluster model and the analysis is an exploratory latent class analysis, and on the other hand, the results in Table 3.2 are those by a confirmatory analysis. In the role conflict for Stouffer-Toby's data set, it may be suitable to assume ordered latent classes that are located in a latent continuum. For the data set, the number of latent classes in the latent class cluster model is less than and equal to three according to the condition of model

Table 3.10 Assignment of the manifest responses to the extracted latent classes based on Table 3.8

θ_1			θ_2				θ_1			θ_2				θ_1			θ_2			
X_{11}	X_{12}	X_{13}	X_{21}	X_{22}	X_{23}	LC[a]	X_{11}	X_{12}	X_{13}	X_{21}	X_{22}	X_{23}	LC	X_{11}	X_{12}	X_{13}	X_{21}	X_{22}	X_{23}	LC
0	0	0	0	0	0	(0,0)	0	1	1	0	1	0	(3,2)	0	0	1	1	0	1	(3,3)
1	0	0	0	0	0	(1,0)	1	1	1	0	1	0	(3,2)	1	0	1	1	0	1	(3,3)
0	1	0	0	0	0	(0,0)	0	0	0	1	1	0	(0,1)	0	1	1	1	0	1	(3,3)
1	1	0	0	0	0	(2,0)	1	0	0	1	1	0	(1,2)	1	1	1	1	0	1	(3,3)
0	0	1	0	0	0	(0,0)	0	1	0	1	1	0	(2,2)	0	0	0	0	1	1	(0,0)
1	0	1	0	0	0	(1,0)	1	1	0	1	1	0	(2,2)	1	0	0	0	1	1	(1,3)
0	1	1	0	0	0	(3,0)	0	0	1	1	1	0	(0,1)	0	1	0	0	1	1	(3,3)
1	1	1	0	0	0	(3,0)	1	0	1	1	1	0	(1,2)	1	1	0	0	1	1	(3,3)
0	0	0	1	0	0	(0,1)	0	1	1	1	1	0	(3,2)	0	0	1	0	1	1	(3,3)
1	0	0	1	0	0	(1,1)	1	1	1	1	1	0	(3,2)	1	0	1	0	1	1	(3,3)
0	1	0	1	0	0	(0,1)	0	0	0	0	0	1	(0,0)	0	1	1	0	1	1	(3,3)
1	1	0	1	0	0	(2,2)	1	0	0	0	0	1	(1,0)	1	1	1	0	1	1	(3,3)
0	0	1	1	0	0	(0,1)	0	1	0	0	0	1	(0,0)	0	0	0	1	1	1	(1,3)
1	0	1	1	0	0	(1,1)	1	1	0	0	0	1	(2,0)	1	0	0	1	1	1	(1,3)
0	1	1	1	0	0	(3,1)	0	0	1	0	0	1	(0,0)	0	1	0	1	1	1	(3,3)

(continued)

Table 3.10 (continued)

θ_1			θ_2				θ_1			θ_2				θ_1			θ_2			
X_{11}	X_{12}	X_{13}	X_{21}	X_{22}	X_{23}	LC[a]	X_{11}	X_{12}	X_{13}	X_{21}	X_{22}	X_{23}	LC	X_{11}	X_{12}	X_{13}	X_{21}	X_{22}	X_{23}	LC
1	1	1	1	0	0	(3,1)	1	0	1	0	0	1	(3,3)	1	1	0	1	1	1	(3,3)
0	0	0	0	1	0	(0,0)	0	1	1	0	0	1	(3,3)	0	0	1	1	1	1	(3,3)
1	0	0	0	1	0	(1,0)	1	1	1	0	0	1	(3,3)	1	0	1	1	1	1	(3,3)
0	1	0	0	1	0	(0,0)	0	0	0	1	0	1	(0,1)	0	1	1	1	1	1	(3,3)
1	1	0	0	1	0	(2,2)	1	0	0	1	0	1	(1,3)	1	1	1	1	1	1	(3,3)
0	0	1	0	1	0	(0,0)	0	1	0	1	0	1	(3,3)							
1	0	1	0	1	0	(1,1)	1	1	0	1	0	1	(3,3)							

[a]LC implies latent class

identification, meanwhile, six in the latent distance model. In order to extract ordered latent classes, it is sensible to make a parsimonious model. Let $\theta_a, a = 1, 2, \ldots, A$ be parameters that express the locations of latent classes in a latent continuum, such that $\theta_a < \theta_{a+1}, a = 1, 2, \ldots, A-1$, and let $\pi_i(\theta_a), i = 1, 2, \ldots, I; a = 1, 2, \ldots, A$ be latent positive response probabilities to binary items i in latent classes a, which satisfy the following inequalities:

$$\pi_i(\theta_a) \le \pi_i(\theta_{a+1}), a = 1, 2, \ldots, A-1; i = 1, 2, \ldots, I. \tag{3.40}$$

The functions $\pi_i(\theta_a)$ are specified before analyzing a data set under study, and it is appropriate that the number of parameters in the model is as small as possible and that the parameters are easy to interpret. Since the positive response probabilities $\pi_i(\theta_a)$ are functions of location or trait parameters θ_a, such models are called structured latent class model. In this section, the following logistic model is used [7]:

$$\pi_i(\theta_\alpha) = \frac{\exp(\theta_\alpha - d_i)}{1 + \exp(\theta_\alpha - d_i)}, a = 1, 2, \ldots A; i = 1, 2, \ldots I, \tag{3.41}$$

where d_i are item difficulty parameters as in the latent trait model and we set $d_1 = 0$ for model identification. The above model is called the Rasch model [15]. Then, the constraints (3.40) are held by the above model. The number of parameters to be estimated is

$$2A + I - 1.$$

Thus, in order to identify the model, we have to keep the following constrain:

$$2A + I - 1 < 2^I - 1 \Leftrightarrow A < \frac{1}{2}(2^I - I), \tag{3.42}$$

because the number of the accounting equations is $2^I - 1$, i.e., the number of the joint probabilities of manifest responses $\boldsymbol{X} = (X_1, X_2, \ldots, X_I)^T$ minus one.

Let $P(\boldsymbol{X} = \boldsymbol{x}|a)$ be the joint probability of $\boldsymbol{X} = (x_1, x_2, \ldots, x_I)^T$ for given latent class a. Then, from (3.41) we have

$$\begin{aligned} P(\boldsymbol{X} = \boldsymbol{x}|a) &= \prod_{i=1}^{I} \left(\frac{\exp(\theta_a - d_i)}{1 + \exp(\theta_a - d_i)}\right)^{x_i} \left(\frac{1}{1 + \exp(\theta_a - d_i)}\right)^{1-x_i} \\ &= \prod_{i=1}^{I} \frac{\exp\{x_i(\theta_a - d_i)\}}{1 + \exp(\theta_a - d_i)}, \end{aligned}$$

$$P(\boldsymbol{X} = \boldsymbol{x}) = \sum_{\alpha=1}^{A} v_a P(\boldsymbol{X} = \boldsymbol{x}|a) = \sum_{\alpha=1}^{A} v_a \prod_{i=1}^{I} \frac{\exp\{x_i(\theta_a - d_i)\}}{1 + \exp(\theta_a - d_i)}.$$

In order to estimate the parameters $\boldsymbol{\phi} = ((v_a), (\theta_a), (d_i))^T$, the EM algorithm is used and the summary of the algorithm is given as follows:

EM algorithm

(i) E-step

Let ${}^s\boldsymbol{\phi} = (({}^s v_a), ({}^s\theta_a), ({}^s d_i))^T$ be the estimate of parameter vector $\boldsymbol{\phi}$ at the s th iteration in the EM algorithm. Then, the conditional expectations of complete data $(n(\boldsymbol{x}, a))$ given parameters ${}^s\boldsymbol{\phi} = (({}^s v_a), ({}^s\theta_a), ({}^s d_i))$ are calculated in the $(s+1)$ th iteration as follows:

$$ {}^{s+1}n(\boldsymbol{x}, a) = n(\boldsymbol{x}) \frac{{}^s v_a \prod_{i=1}^{I} {}^s P(X_i = x_i | a)}{\sum_{b=0}^{I} {}^s v_b \prod_{i=1}^{I} {}^s P(X_i = x_i | b)}, a = 0, 1, 2, \ldots, I, \quad (3.43) $$

where

$$ {}^s P(X_i = x_i | a) = \frac{\exp\{x_i({}^s\theta_a - {}^s d_i)\}}{1 + \exp({}^s\theta_a - {}^s d_i)}, x_i = 0, 1. $$

(ii) M-step

As in (3.10), the loglikelihood function of the complete data $\left({}^{s+1}n(\boldsymbol{x}, a)\right)$ (3.43) is given by

$$ Q(\boldsymbol{\phi}|{}^s\boldsymbol{\phi}) = l\left(\boldsymbol{\phi}|\left({}^{s+1}n(\boldsymbol{x}, a)\right)\right) $$
$$ = \sum_{\alpha=1}^{A} \sum_{\boldsymbol{x}} {}^{s+1}n(\boldsymbol{x}, a) \log\left(v_a \prod_{i=1}^{I} \frac{\exp\{x_i(\theta_a - d_i)\}}{1 + \exp(\theta_a - d_i)} \right) $$
$$ = \sum_{\alpha=1}^{A} \sum_{\boldsymbol{x}} {}^{s+1}n(\boldsymbol{x}, a) \log v_a $$
$$ + \sum_{\alpha=1}^{A} \sum_{\boldsymbol{x}} {}^{s+1}n(\boldsymbol{x}, a) \left[\sum_{i=1}^{I} \{x_i(\theta_a - d_i) - \log(1 + \exp(\theta_a - d_i))\} \right]. \quad (3.44) $$

With respect to ${}^{s+1}v_a$, we have

$$ {}^{s+1}v_a = \frac{\sum_{\boldsymbol{x}} {}^{s+1}n(\boldsymbol{x}, a)}{N}, a = 1, 2, \ldots, A; $$

however, for estimating the other parameters θ_a and d_i, the Newton-Raphson method needs to be used in the M-step. The first derivatives of $Q(\boldsymbol{\phi}|{}^s\boldsymbol{\phi})$ with respect to θ_a and d_i, respectively, are calculated as follows:

$$\frac{\partial Q(\boldsymbol{\phi}|{}^{s}\boldsymbol{\phi})}{\partial \theta_a} = \sum_{i=1}^{I} \sum_{\boldsymbol{x}} {}^{s+1}n(\boldsymbol{x}, a)(x_i - P(X_i = 1|a)), a = 1, 2, \ldots, A;$$

$$\frac{\partial Q(\boldsymbol{\phi}|{}^{s}\boldsymbol{\phi})}{\partial d_i} = -\sum_{\alpha=1}^{A} \sum_{\boldsymbol{x}} {}^{s+1}n(\boldsymbol{x}, a)(x_i - P(X_i = 1|a)), i = 2, 3, \ldots, I.$$

Then, the $(A + I)$-dimensional gradient vector is set as

$$\boldsymbol{g} = \left(\begin{pmatrix} \frac{\partial Q(\boldsymbol{\phi}|{}^{s}\boldsymbol{\phi})}{\partial \theta_a} \end{pmatrix} \\ \begin{pmatrix} \frac{\partial Q(\boldsymbol{\phi}|{}^{s}\boldsymbol{\phi})}{\partial d_i} \end{pmatrix} \right). \tag{3.45}$$

Consequently, the second partial derivatives of $Q(\boldsymbol{\phi}|{}^{s}\boldsymbol{\phi})$ are calculated as follows:

$$\frac{\partial^2 Q(\boldsymbol{\phi}|{}^{s}\boldsymbol{\phi})}{\partial \theta_a^2} = -\sum_{\boldsymbol{x}} {}^{s+1}n(\boldsymbol{x}, a) \sum_{i=1}^{I} P(X_i = 1|a)(1 - P(X_i = 1|a))$$
$$= -N \cdot {}^{s+1}v_a \sum_{i=1}^{I} P(X_i = 1|a)(1 - P(X_i = 1|a)),$$
$$a = 1, 2, \ldots, A;$$

$$\frac{\partial^2 Q(\boldsymbol{\phi}|{}^{s}\boldsymbol{\phi})}{\partial \theta_a \partial d_i} = \sum_{\boldsymbol{x}} {}^{s+1}n(\boldsymbol{x}, a) P(X_i = 1|a)(1 - P(X_i = 1|a))$$
$$= N \cdot {}^{s+1}v_a P(X_i = 1|a)(1 - P(X_i = 1|a)),$$
$$a = 1, 2, \ldots, A; i = 1, 2, \ldots, I;$$

$$\frac{\partial^2 Q(\boldsymbol{\phi}|{}^{s}\boldsymbol{\phi})}{\partial d_i^2} = -\sum_{\alpha=1}^{A} \sum_{\boldsymbol{x}} {}^{s+1}n(\boldsymbol{x}, a) P(X_i = 1|a)(1 - P(X_i = 1|a))$$
$$= -\sum_{\alpha=1}^{A} N \cdot {}^{s+1}v_a P(X_i = 1|a)(1 - P(X_i = 1|a)),$$
$$i = 2, 3, \ldots, I.$$

$$\frac{\partial^2 Q(\boldsymbol{\phi}|{}^{s}\boldsymbol{\phi})}{\partial \theta_a \partial \theta_b} = \frac{\partial^2 Q(\boldsymbol{\phi}|{}^{s}\boldsymbol{\phi})}{\partial d_i \partial d_j} = 0, a \neq b; i \neq j.$$

From the above results, the Hessian matrix $\boldsymbol{H}$ is set as follows:

Table 3.11 The estimated latent ordered-class model with five latent classes (Stouffer-Toby's data)

Latent class θ	Proportion	Latent positive item response probability			
		X_1	X_2	X_3	X_4
1(−1.563)	0.058	0.173	0.037	0.040	0.011
2(1.072)	0.124	0.745	0.348	0.365	0.133
3(1.244)	0.267	0.776	0.388	0.406	0.154
4(1.329)	0.320	0.791	0.408	0.427	0.166
5(4.771)	0.231	0.992	0.956	0.959	0.861

$G^2 = 1.092(df = 3, P = 0.779)$

$$\boldsymbol{H} = \begin{pmatrix} \left(\frac{\partial^2 Q(\boldsymbol{\phi}|{}^s\boldsymbol{\phi})}{\partial\theta_a \partial\theta_b}\right) & \left(\frac{\partial^2 Q(\boldsymbol{\phi}|{}^s\boldsymbol{\phi})}{\partial\theta_a \partial d_i}\right) \\ \left(\frac{\partial^2 Q(\boldsymbol{\phi}|{}^s\boldsymbol{\phi})}{\partial d_i \partial\theta_a}\right) & \left(\frac{\partial^2 Q(\boldsymbol{\phi}|{}^s\boldsymbol{\phi})}{\partial d_i \partial d_j}\right) \end{pmatrix}. \tag{3.46}$$

Let $\boldsymbol{\phi}_{(\theta,d)} = ((\theta_a), (d_i))$ and let the t th iterative value of $\boldsymbol{\phi}_{(\theta,d)}$ be $\boldsymbol{\phi}_{(\theta,d)t} = ((\theta_{at}), (d_{it}))$, where $\boldsymbol{\phi}_{(\theta,d)1} = (({}^s\theta_a), ({}^s d_i))$. Then, $\boldsymbol{\phi}_{(\alpha,\beta)t+1}$ is obtained as follows:

$$\boldsymbol{\phi}_{(\theta,d)t+1} = \boldsymbol{\phi}_{(\theta,d)t} - \boldsymbol{H}_t^{-1}\boldsymbol{g}_t, t = 1, 2, \ldots,$$

where $\boldsymbol{H}_t$ and $\boldsymbol{g}_t$ are values of the gradient vector (3.45) and the Hessian matrix (3.46) at $\boldsymbol{\phi}_t = \left(\left({}^{s+1}v_a\right), \boldsymbol{\phi}_{(\theta,d)t}\right)$. From this algorithm, we can get $\lim_{t\to\infty}\boldsymbol{\phi}_{(\theta,d)t} = \left(\left({}^{s+1}\theta_a\right), \left({}^{s+1}d_i\right)\right)$.

The above model is applied to the analysis of Stouffer-Toby's data in Table 2.1. Considering the inequality constraint (3.42), first, for the number of ordered latent classes $A = 5$, the results are shown in Table 3.11. The goodness-of-fit to the data set is good as indicated in the table, $G^2 = 1.092(df = 3, P = 0.779)$. According to the latent response probabilities to four items in latent classes 2 through 4 (Fig. 3.6), the latent classes are similar, and second, the latent ordered-class model with three latent classes is used for analyzing the data. The results are illustrated in Table 3.12, and it shows the goodness-of-fit to the data is very good, $G^2 = 1.092(df = 7, P = 0.993)$. The response probabilities for three latent classes are illustrated in Fig. 3.7. In Sect. 2.3, the data have been analyzed with usual latent class models with three and two latent classes. Although the ordered latent classes have been derived as shown in Tables 2.2 and 2.3, the present models are more parsimonious than the usual models. The assessment of respondents based on the latent ordered-class analysis in Table 3.12 is demonstrated in Table 3.13.

In this section, latent ordered-class analysis is discussed by using model (3.41); however, the analysis does not confine us to the model. Structured models can be made as far as the models are identified. The present approach assumes ordered latent classes in the population, which can be located in a latent continuum. In this sense, it is related to item response models that assume latent continuums in the populations.

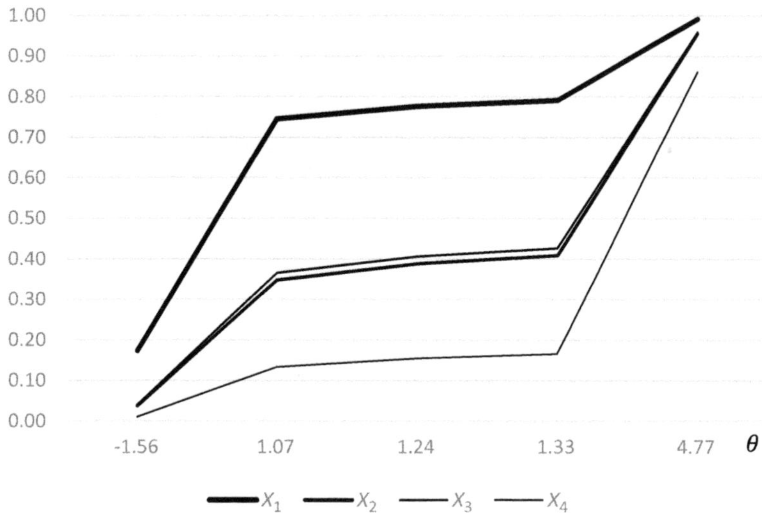

Fig. 3.6 Locations of five ordered latent classes and their positive response probabilities in Table 3.11

Table 3.12 The estimated latent ordered-class model with three latent classes (Stouffer-Toby's data)

Latent class	Proportion	Latent positive item response probability			
		X_1	X_2	X_3	X_4
1(−0.397)	0.150	0.434	0.123	0.131	0.039
2(1.412)	0.647	0.811	0.439	0.458	0.184
3(5.160)	0.203	0.995	0.974	0.975	0.914

$G^2 = 1.092(df = 7,\ P = 0.993)$

The explanatory power of latent ordered-class model (3.41) is assessed by the entropy coefficient determination (ECD) (1.30). In this model, the ECD is calculated as

$$\mathrm{ECD}(X, \theta) = \frac{\sum_{i=1}^{I} \mathrm{Cov}(X_i, \theta)}{1 + \sum_{i=1}^{I} \mathrm{Cov}(X_i, \theta)}, \tag{3.47}$$

where

$$\mathrm{Cov}(X_i, \theta) = \sum_{a=1}^{A} v_a \pi_i(\theta_a)\theta_a - \mathrm{E}(X_i)\mathrm{E}(\theta)$$

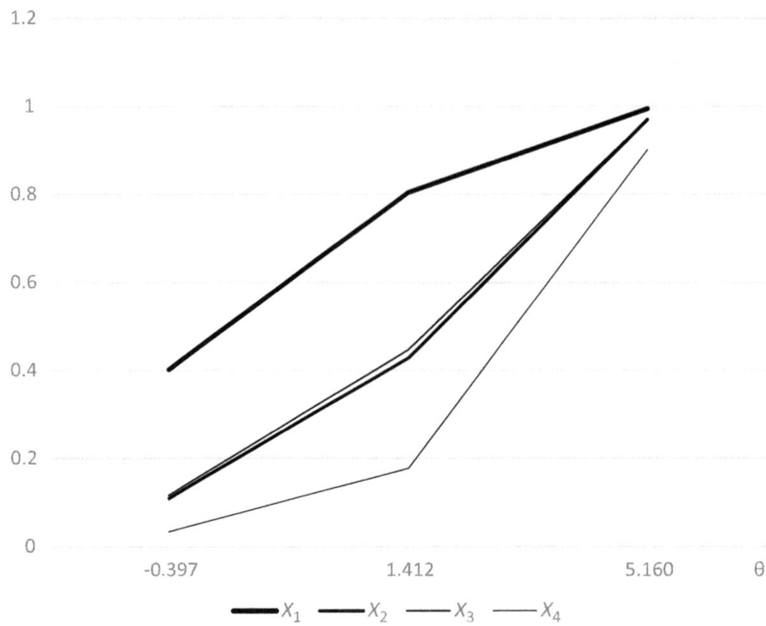

Fig. 3.7 Location of three latent ordered classes and their positive response probabilities in Table 3.12

Table 3.13 Assignment of the manifest responses to the extracted latent classes (latent ordered-class analysis of data in Table 2.1)

Response pattern	Score	Latent class	Response pattern	Score	Latent class
0 0 0 0	0	1	0 0 0 1	1	2
1 0 0 0	1	2	1 0 0 1	2	2
0 1 0 0	1	2	0 1 0 1	2	2
1 1 0 0	2	2	1 1 0 1	3	2
0 0 1 0	1	2	0 0 1 1	2	2
1 0 1 0	2	2	1 0 1 1	3	2
0 1 1 0	2	2	0 1 1 1	3	2
1 1 1 0	3	2	1 1 1 1	4	3

$$= \sum_{a=1}^{A} v_a \pi_i(\theta_a)\theta_a - P(X_i = 1) \sum_{a=1}^{A} v_a \theta_a,$$

$$i = 1, 2, \ldots, I.$$

From this, we can obtain ECD (3.47) and ECDs for manifest variables X_i are also given by

Table 3.14 The explained entropy $\mathrm{Cov}(X_i, \theta)$ and ECDs in the estimated model in Table 3.12

Manifest variable	$\mathrm{Cov}(X_i, \theta)$	ECD
X_1	0.250	0.200
X_2	0.459	0.315
X_3	0.451	0.311
X_4	0.528	0.345
Total	1.688	0.628

$$\mathrm{ECD}(X_i, \theta) = \frac{\mathrm{Cov}(X_i, \theta)}{1 + \mathrm{Cov}(X_i, \theta)}. \tag{3.48}$$

By using (3.47) and (3.48), the estimated model in Table 3.12 is assessed. The results are demonstrated in Table 3.14. Since $\mathrm{ECD}(\boldsymbol{X}, \theta) = 0.628$, 62.8% of the uncertainty of manifest variable vector $\boldsymbol{X} = (X_1, X_2, X_3, X_4)^T$ is explained by the model.

In this section, latent ordered-class model (3.41) has been discussed and the model has been applied to data in Table 2.1. A more general model is given by

$$\pi_i(\theta_a) = \frac{\exp(\beta_i(\theta_a - d_i))}{1 + \exp(\beta_i(\theta_a - d_i))}, a = 1, 2, \ldots, A; i = 1, 2, \ldots, I, \tag{3.49}$$

where parameters β_i indicate regression parameters as in the two-parameter item response model. Then, in order to identify the model, for example, we set $\beta_1 = 1, d_1 = 0$, and the following inequality has to be held:

$$2^I - 2(A + I - 1) > 0.$$

The ECDs of two-parameter model (3.49) are given by

$$\mathrm{ECD}(\boldsymbol{X}, \theta) = \frac{\sum_{i=1}^{I} \beta_i \mathrm{Cov}(X_i, \theta)}{1 + \sum_{i=1}^{I} \beta_i \mathrm{Cov}(X_i, \theta)},$$

and

$$\mathrm{ECD}(X_i, \theta) = \frac{\beta_i \mathrm{Cov}(X_i, \theta)}{1 + \beta_i \mathrm{Cov}(X_i, \theta)}, i = 1, 2, \ldots, I.$$

3.6 The Latent Trait Model (Item Response Model)

In this section, the ML estimation of two-parameter logistic model (1.10) is considered. As in Sect. 1.3 in Chap. 1, the model is approximated by a latent class model (1.9). As the positive response probabilities in the latent class model are equivalent to those in (3.49), the discussion below will be made with (3.49), where $\beta_i = Da_i, i = 1, 2, \ldots, I$. In this case, parameters to be estimated are discriminant parameters β_i and item difficulties $d_i, i = 1, 2, \ldots, I$, whereas the class proportions v_a are given by the standard normal distribution and latent trait parameters $\theta_a, a = 1, 2, \ldots, A$ are also given for the approximation (1.7). For an appropriate division of the latent continuum θ given in (1.7), we calculate class proportions v_a by (1.8). Then, the latent class model is set as

$$P(\boldsymbol{X} = \boldsymbol{x}) = \sum_{\alpha=1}^{A} v_a \prod_{i=1}^{I} P(X_i = x_i|\theta_a) = \sum_{\alpha=1}^{A} v_a \prod_{i=1}^{I} \frac{\exp(x_i \beta_i(\theta_a - d_i))}{1 + \exp(\beta_i(\theta_a - d_i))}, \tag{3.50}$$

and the EM algorithm for the ML estimation is given as follows:

EM algorithm

(i) E-step

Let ${}^s\boldsymbol{\phi} = (({}^s\beta_i), ({}^s d_i))^{\boldsymbol{T}}$ be the estimate of parameter vector $\boldsymbol{\phi}$ at the s th iteration in the EM algorithm. Then, the conditional expectations of complete data $(n(\boldsymbol{x}, a))$ for given parameters ${}^s\boldsymbol{\phi} = (({}^s\beta_i), ({}^s d_i))$ are calculated in the $(s + 1)$ th iteration as follows:

$${}^{s+1}n(\boldsymbol{x}, a) = n(\boldsymbol{x}) \frac{v_a \prod_{i=1}^{I} {}^s P(X_i = x_i|\theta_a)}{\sum_{b=0}^{I} v_a \prod_{i=1}^{I} {}^s P(X_i = x_i|\theta_b)}, a = 0, 1, 2, \ldots, I, \tag{3.51}$$

where

$${}^s P(X_i = x_i|\theta_a) = \frac{\exp(x_i {}^s\beta_i(\theta_a - {}^s d_i))}{1 + \exp({}^s\beta_i(\theta_a - {}^s d_i))}, x_i = 0, 1.$$

(ii) M-step

The log likelihood function of the complete data $\left({}^{s+1}n(\boldsymbol{x}, a)\right)$ (3.51) is given by

$$\begin{aligned} Q(\boldsymbol{\phi}|{}^s\boldsymbol{\phi}) &= l\left(\boldsymbol{\phi}|\left({}^{s+1}n(\boldsymbol{x}, a)\right)\right) \\ &= \sum_{\alpha=1}^{A} \sum_{\boldsymbol{x}} {}^{s+1}n(\boldsymbol{x}, a) \log\left(v_a \prod_{i=1}^{I} \frac{\exp(x_i \beta_i(\theta_a - d_i))}{1 + \exp(\beta_i(\theta_a - d_i))} \right) \end{aligned}$$

$$= \sum_{\alpha=1}^{A} \sum_{\boldsymbol{x}} {}^{s+1}n(\boldsymbol{x}, a)\log v_a$$

$$+ \sum_{\alpha=1}^{A} \sum_{\boldsymbol{x}} {}^{s+1}n(\boldsymbol{x}, a)\left[\sum_{i=1}^{I}\{x_i\beta_i(\theta_a - d_i) - \log(1 + \exp(\beta_i(\theta_a - d_i)))\}\right]. \quad (3.52)$$

For estimating the other parameters β_i and d_i, the Newton-Raphson method needs to be used in the M-step. The first derivatives of $Q(\boldsymbol{\phi}|^s\boldsymbol{\phi})$ with respect to θ_a and d_i, respectively, are calculated as follows:

$$\frac{\partial Q(\boldsymbol{\phi}|^s\boldsymbol{\phi})}{\partial \beta_i} = \sum_{a=1}^{A} \sum_{\boldsymbol{x}} {}^{s+1}n(\boldsymbol{x}, a)(\theta_a - d_i)(x_i - P(X_i = x_i|\theta_a)), i = 1, 2, \ldots, I; \quad (3.53)$$

$$\frac{\partial Q(\boldsymbol{\phi}|^s\boldsymbol{\phi})}{\partial d_i} = -\sum_{\alpha=1}^{A} \sum_{\boldsymbol{x}} {}^{s+1}n(\boldsymbol{x}, a)\beta_i(x_i - P(X_i = x_i|\theta_a)), i = 1, 2, \ldots, I. \quad (3.54)$$

Then, the $2I$-dimensional gradient vector is set as

$$\boldsymbol{g} = \left(\begin{pmatrix} \frac{\partial Q(\boldsymbol{\phi}|^s\boldsymbol{\phi})}{\partial \beta_i} \\ \frac{\partial Q(\boldsymbol{\phi}|^s\boldsymbol{\phi})}{\partial d_i} \end{pmatrix} \right). \quad (3.55)$$

Consequently, the second-order partial derivatives of $Q(\boldsymbol{\phi}|^s\boldsymbol{\phi})$ are calculated as follows:

$$\frac{\partial^2 Q(\boldsymbol{\phi}|^s\boldsymbol{\phi})}{\partial \beta_i^2} = -\sum_{a=1}^{A} \sum_{\boldsymbol{x}} {}^{s+1}n(\boldsymbol{x}, a)(\theta_a - d_i)^2 P(X_i = 1|a)(1 - P(X_i = 1|a)), \quad (3.56)$$

$$a = 1, 2, \ldots, A;$$

$$\frac{\partial^2 Q(\boldsymbol{\phi}|^s\boldsymbol{\phi})}{\partial \beta_i \partial d_i} = -\sum_{a=1}^{A} \sum_{\boldsymbol{x}} {}^{s+1}n(\boldsymbol{x}, a)$$
$$\{(x_i - P(X_i = 1|a)) - (\theta_a - d_i)\beta_i P(X_i = 1|a)(1 - P(X_i = 1|a))\}, \quad (3.57)$$

$$i = 1, 2, \ldots, I;$$

$$\frac{\partial^2 Q(\boldsymbol{\phi}|{}^s\boldsymbol{\phi})}{\partial d_i^2} = -\sum_{\alpha=1}^{A}\sum_{\boldsymbol{x}} {}^{s+1}n(\boldsymbol{x}, a)\beta_i^2 P(X_i = 1|a)(1 - P(X_i = 1|a)), \quad (3.58)$$

$$i = 2, 3, \ldots, I.$$

$$\frac{\partial^2 Q(\boldsymbol{\phi}|{}^s\boldsymbol{\phi})}{\partial \beta_i \partial \beta_j} = \frac{\partial^2 Q(\boldsymbol{\phi}|{}^s\boldsymbol{\phi})}{\partial d_i \partial d_j} = 0, i \neq j. \quad (3.59)$$

From the above results, the Hessian matrix $\boldsymbol{H}$ is set as follows:

$$\boldsymbol{H} = \begin{pmatrix} \left(\frac{\partial^2 Q(\boldsymbol{\phi}|{}^s\boldsymbol{\phi})}{\partial \beta_i \partial \beta_j}\right) & \left(\frac{\partial^2 Q(\boldsymbol{\phi}|{}^s\boldsymbol{\phi})}{\partial \beta_i \partial d_i}\right) \\ \left(\frac{\partial^2 Q(\boldsymbol{\phi}|{}^s\boldsymbol{\phi})}{\partial d_i \partial \beta_i}\right) & \left(\frac{\partial^2 Q(\boldsymbol{\phi}|{}^s\boldsymbol{\phi})}{\partial d_i \partial d_j}\right) \end{pmatrix}. \quad (3.60)$$

Let ${}^s\boldsymbol{\phi}_t = (({}^s\beta_{it}), ({}^s d_{it}))$ be the t th iterative value of $\boldsymbol{\phi}$, where ${}^s\boldsymbol{\phi}_1 = (({}^s\beta_i), ({}^s d_i))$. Then, ${}^s\boldsymbol{\phi}_{t+1}$ is obtained as follows:

$${}^s\boldsymbol{\phi}_{t+1} = {}^s\boldsymbol{\phi}_t - \boldsymbol{H}_t^{-1}\boldsymbol{g}_t, t = 1, 2, \ldots,$$

where $\boldsymbol{H}_t$ and $\boldsymbol{g}_t$ are values of the gradient vector (3.55) and the Hessian matrix (3.60) at ${}^s\boldsymbol{\phi}_t$.

Remark 3.6 The expectation of the Hessian matrix is minus the Fisher information matrix. Although the Fisher information matrix is positive definite, however, Hessian matrices (3.60) calculated in the iterations are not necessarily guaranteed to be negative definite. Since in latent class a,

$$\mathrm{E}\{X_i - P(X_i = 1|a)|a\} = 0,$$

$$E\left\{(X_i - P(X_i = 1|a))^2|a\right\} = P(X_i = 1|a)(1 - P(X_i = 1|a)), i = 1, 2, \ldots, I,$$

for large samples, we can use the following approximation of (3.57):

$$\frac{\partial^2 Q(\boldsymbol{\phi}|{}^s\boldsymbol{\phi})}{\partial \beta_i \partial d_i} \approx \sum_{a=1}^{A}\sum_{\boldsymbol{x}} {}^{s+1}n(\boldsymbol{x}, a)(\theta_a - d_i)\beta_i P(X_i = 1|a)(1 - P(X_i = 1|a)), \quad (3.61)$$

$$i = 1, 2, \ldots, I;$$

Then, the Hessian matrix is always negative definite. ■

First, latent class model (3.50) is applied to estimate the latent trait model by using Stouffer-Toby's data in Table 2.1. Before analyzing the data, latent continuous trait

Table 3.15 Upper limits of latent classes $\theta_{(a)}$, class values θ_a and latent class proportions v_a

a	1	2	3	4	5	6	7	8	9	10
$\theta_{(a)}$	−2	−1.5	−1	−0.5	0	0.5	1	1.5	2	$+\infty$
θ_a	−2.5	−1.75	−1.25	−0.75	−0.25	0.25	0.75	1.25	1.75	2.5
v_a	0.023	0.044	0.092	0.150	0.191	0.191	0.150	0.092	0.044	0.023

Table 3.16 The estimated parameters in latent trait model (3.50) from the Stouffer-Toby's data

Manifest variable	X_1	X_2	X_3	X_4
β_i	1.128	1.559	1.330	2.076
d_i	−1.471	−0.006	−0.061	0.643

$G^2 = 8.570, (df = 7, P = 0.285)$

θ is divided as in (1.7). In order to demonstrate the above method, latent trait θ is divided into ten intervals (latent classes):

$$-\infty\left(=\theta_{(0)}\right) < \theta_{(1)} < \theta_{(2)} < \cdots < \theta_{(9)} < +\infty\left(=\theta_{(10)}\right), \tag{3.62}$$

and the class values

$$\theta_{(a-1)} < \theta_a \le \theta_{(a)}, a = 1, 2, \ldots, 10$$

are set. The values and the latent class proportions v_a calculated with the standard normal distribution (1.8) are illustrated in Table 3.15. These values are fixed in the estimation procedure. In the estimation procedure, (3.61) is employed for (3.57), and the estimated parameters are given in Table 3.16. According to the test of goodness-of-fit, latent trait model (3.58) fairly fits the data set and it is reasonable to assume a latent continuous trait to respond to the four test items. The graphs of the item response functions are illustrated in Fig. 3.8. The assessment of test items (manifest variables) as indicators of the latent trait is shown in Table 3.17. In order to estimate latent trait θ of an individual with response vector $\boldsymbol{x}$, the following method is used. Let $f(\boldsymbol{x}, \theta)$ be the joint probability function of $\boldsymbol{x} = (x_1, x_2, x_3, x_4)$ and θ. The estimate is given by θ_{max} such that $f(\boldsymbol{x}, \theta_{max}) = \max_{\theta} f(\boldsymbol{x}, \theta)$. Since θ is distributed according to the standard normal distribution $\varphi(\theta)$, from (1.6) we have

$$\log f(\boldsymbol{x}, \theta) = -\frac{1}{2}\log 2\pi - \frac{\theta^2}{2} + \log \prod_{i=1}^{I} \frac{\exp(x_i \beta_i(\theta - d_i))}{1 + \exp(\beta_i(\theta - d_i))}$$

$$= -\frac{1}{2}\log 2\pi - \frac{\theta^2}{2} + \sum_{i=1}^{I} x_i \beta_i(\theta - d_i) + \sum_{i=1}^{I} \log(1 + \exp(\beta_i(\theta - d_i))).$$

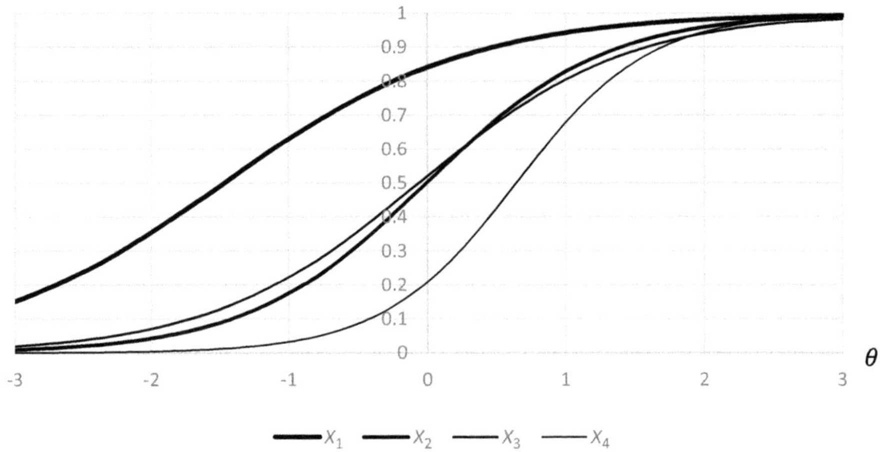

Fig. 3.8 Graph of latent trait model (3.50) estimated from Stouffer-Toby's data (Table 2.1)

Table 3.17 The explained entropy $\mathrm{Cov}(X_i, \theta)$ and ECDs in the estimated model in Table 3.15

Manifest variable	$\beta_i \mathrm{Cov}(X_i, \theta)$	ECD
X_1	0.155	0.134
X_2	0.271	0.213
X_3	0.248	0.199
X_4	0.271	0.213
Total	0.946	0.486

In order to maximize the above function with respect to θ, differentiating the above function with respect to θ and setting it to zero, we obtain

$$\frac{d}{d\theta}\log f(\boldsymbol{x}, \theta) = -\theta + \sum_{i=1}^{I} \beta_i (x_i - P(X_i = 1|\theta)) = 0. \tag{3.63}$$

By solving the above equation with respect to θ, we can get the estimate of latent trait θ of a respondent with manifest response vector $\boldsymbol{x} = (x_1, x_2, x_3, x_4)$ (Table 3.18).

Remark 3.7 In order to solve Eq. (3.63), the following Newton–Raphson method is employed. Let $\theta^{(m)}$ be the estimate of θ in the m th iteration. Then, the algorithm for obtaining a solution in (3.63) is given by

$$\theta^{(m+1)} = \theta^{(m)} - \frac{\frac{d}{d\theta}\log f\left(\boldsymbol{x}, \theta^{(m)}\right)}{\frac{d^2}{d\theta^2}\log f\left(\boldsymbol{x}, \theta^{(m)}\right)}, m = 0, 1, 2, \ldots,$$

where

Table 3.18 Assessment of respondents by using the estimated latent trait model (Table 3.15)

Response pattern	θ_a	Response pattern	θ_a
0 0 0 0	−1.097	0 0 0 1	−0.320
1 0 0 0	−0.515	1 0 0 1	−0.150
0 1 0 0	−0.836	0 1 0 1	0.278
1 1 0 0	0.289	1 1 0 1	0.384
0 0 1 0	−0.313	0 0 1 1	0.051
1 0 1 0	0.061	1 0 1 1	0.157
0 1 1 0	−0.464	0 1 1 1	0.585
1 1 1 0	0.658	1 1 1 1	0.900

Table 3.19 The estimated parameters in latent trait model (3.50) from the Lazarsfeld-Stouffer's data

Manifest variable	X_1	X_2	X_3	X_4
β_i	1.672	1.099	1.391	1.577
d_i	0.525	−0.586	−0.831	−0.983

$G^2 = 7.515$, $(df = 7,\ P = 0.377)$

$$\frac{d^2}{d\theta^2}\log f(\boldsymbol{x}, \theta) = -1 - \sum_{i=1}^{I} \beta_i^2 P(X_i = 1|\theta)(1 - P(X_i = 1|\theta)).$$

Second, McHugh's data in Table 2.5 are analyzed with model (3.50). The log likelihood ratio test statistic $G^2 = 22.011(df = 7,\ P = 0.003)$ is obtained, and thus, the model fitness to the data set is bad. It may be concluded that there is no latent continuous trait distributed according to the standard normal distribution or the latent trait space not one-dimensional. Finally, Lazarsfel-Stouffer's data (Table 2.7) are analyzed with model (3.50), and the estimated parameters and the latent response probabilities $P(X_i = x_i|\theta)$ are illustrated in Table 3.19 and Fig. 3.9, respectively. The latent trait model makes a moderate fit to the data set, that is, $G^2 = 7.515(df = 7,\ P = 0.377)$. The predictive or explanatory power of latent trait θ for manifest variables (Table 3.20) is similar to that of Stouffer-Toby's data (Table 3.16). As demonstrated above, the latent trait model can be estimated in a framework of the latent class model, and the EM algorithm is effective to estimate the model parameters.

3.7 Discussion

In this chapter, latent class analyses with ordered latent classes have been discussed. In latent distance analysis, the model is an extension of the Guttman scale model, and the intrusion and omission errors are incorporated into the model itself. Assuming a

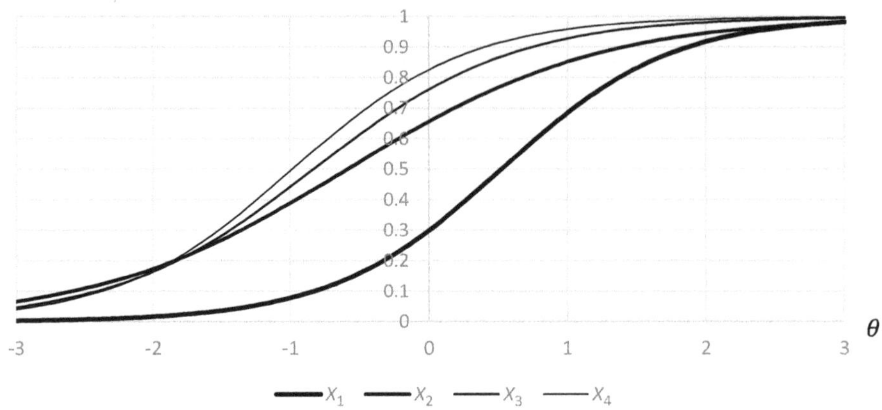

Fig. 3.9 Graph of latent trait model (3.50) estimated from Lazarsfeld-Stouffer's data (Table 2.7)

Table 3.20 The explained entropy $\mathrm{Cov}(X_i, \theta)$ and ECDs in the estimated model in Table 3.17

Manifest variable	$\beta_i \mathrm{Cov}(X_i, \theta)$	ECD
X_1	0.258	0.205
X_2	0.208	0.172
X_3	0.219	0.180
X_4	0.215	0.177
Total	0.901	0.474

latent one-dimensional continuum, the positive response probabilities are structured with threshold parameters to respond positively (successfully) to items. Another model is constructed with a logit model with location parameters. The location parameters are introduced to assess the levels of latent classes in one-dimensional continuum, for example, trait, ability, and so on. In this sense, the model is viewed as a discrete version of the latent trait model, that is, the Rash model. In the present chapter, a latent trait model with discriminant parameters and item difficulties, that is, a two-parameter logistic model, is also treated, and a latent class model approach to the ML estimation of the parameters is provided, i.e., the ML estimation procedure based on the EM algorithm. The method is demonstrated by using data sets in Chapter 2. The latent class models in this chapter can deal with more latent classes than latent class cluster model in Chapter 2. In practical data analyses, it is effective to use the latent class model that incorporates ordered latent classes into the model itself, as demonstrated in this chapter. Moreover, it is sensible to challenge to make new latent class models flexibly for purposes of data analyses, and it leads to a development of latent class analysis.

References

1. Croon, M. A. (1990). Latent class analysis with ordered latent classes. *British Journal of Mathematical and Statistical Psychology, 43*, 171–192.
2. Dayton, C. M., & Macready, G. B. (1976). A probabilistic model for validation of behavioral hierarchies. *Psychometrika, 43*, 189–204.
3. Dayton, C. M., & Macready, G. B. (1980). A scaling model with response errors and intrinsically unscalable responses. *Psychometrika, 45*, 343–356.
4. De Leeuw, J., & Verhelst, N. (1986). Maximum likelihood estimation in generalized Rasch models. *Journal of Educational Statistics, 11*, 183–196.
5. Dempster, A. P., Laird, N. M., & Rubin, D. B. (1977). Maximum likelihood from incomplete data via the EM algorithm (with discussion). *J R Stat Soc B, 39*, 1–38.
6. Eshima, N., & Asano, C. (1988). On latent distance analysis and the MLE algorithm. *Behaviormetrika, 24*, 25–32.
7. Eshima, N., & Asano, C. (1989). Latent ordered class analysis. *Bull Comput Stat Jpn, 2*, 25–34. (in Japanese).
8. Eshima, N. (1992). A hierarchical assessment of latent traits by using latent Guttman scaling. *Behaviormetrika, 19*, 97–116.
9. Lazarsfeld, P. F., & Henry, N. M. (1968). *Latent structure analysis*. Boston: Houghton Mifflin.
10. Lindsay, B., Clogg, C., & Grego, J. (1991). Semiparametric estimation in the Rasch model and related exponential response models, including a simple latent class model for item analysis. *Journal of American Statistical Association, 86*, 96–107.
11. Magidson, J., & Vermunt, J. K. (2001). Latent class factor and cluster models: Bi-plots, and related graphical displays. *Sociological Methodology, 31*, 223–264.
12. McHugh, R. B. (1956). Efficient estimation of local identification in latent class analysis. *Psychometrika, 20*, 331–347.
13. Price, L. C., Dayton, C. M., & Macready, G. B. (1980). Discovery algorithms for hierarchical relations. *Psychometrika, 45*, 449–465.
14. Proctor, C. H. (1970). A probabilistic formulation and statistical analysis of Guttman scaling. *Psychometrika, 35*, 73–78.
15. Rasch, G. (1980). *Probabilistic models for some intelligence and attainment tests*. Illinois: The University of Chicago Press.
16. Stouffer SA, Toby J (1951) Role conflict and personality. Am J Soc 56:395–406
17. Vermunt, J. K. (2010). Latent class models. *Int Encycl Educ, 7*, 238–244.

Chapter 4
Latent Class Analysis with Latent Binary Variables: An Application for Analyzing Learning Structures

4.1 Introduction

Usual latent class analysis is carried out without any assumptions on latent response probabilities for test items. In this sense, latent classes in the analysis are treated parallelly and the analysis is referred to as the latent class cluster analysis [10]. In Chap. 3, latent class analyses with ordered latent classes have been discussed with models that incorporate the ordered structures into the models themselves. In latent distance analysis, the response items are ordered with respect to the item levels (difficulties) that are located in a one-dimensional latent continuum, and an individual beyond the levels responds to the correspondent items with higher probabilities than an individual with below the levels. The latent distance model can be applied to learning studies as well, for example, assessing individuals' acquisition states of several skills for solving test binary items. Let $X_i, i = 1, 2, \ldots, I$ be manifest response variables corresponding to items i, such that

$$X_i = \begin{cases} 1 \ (success\ to\ item\ i) \\ 0 \qquad (failure) \end{cases}, \tag{4.1}$$

and let $S_i, i = 1, 2, \ldots, I$ be acquisition states of skills i for solving test items i, such that

$$S_i = \begin{cases} 1 \ (acquisition\ of\ skill\ i) \\ 0 \quad (non - acquisition) \end{cases}. \tag{4.2}$$

In this case, the test scales the states of the skill acquisitions, which are not observed directly, and thus, S_i are viewed as latent binary variables. In the above assumption, the following inequalities for success probabilities for test items are naturally required:

N. Eshima, *An Introduction to Latent Class Analysis*, Behaviormetrics: Quantitative Approaches to Human Behavior 14,
https://doi.org/10.1007/978-981-19-0972-6_4

$$P(X_i = 1|S_i = 1) > P(X_i = 1|S_i = 0), i = 1, 2, \ldots, I.$$

If the skills under study have prerequisite relations, for example, skill i prerequisite to skill $i+1$, $i = 1, 2, \ldots, I$, then, the latent states of skill acquisition are

$$(S_1, S_2, \ldots, S_I) = (0, 0, \ldots, 0), (1, 0, \ldots, 0), \ldots, (1, 1, \ldots, 1)$$

and the latent states correspond to latent classes, so the model is the same as the latent distance model. For example, assuming S_1 is the state of addition skill in arithmetic, S_2 that of multiplication skill and S_3 that of division skill, then, the skill of addition is prerequisite to that of multiplication and the skill of multiplication is prerequisite to that of division, and the scale patterns (S_1, S_2, S_3) are $(0, 0, 0)$, $(1, 0, 0)$, $(1, 1, 0)$, and$(1, 1, 1)$. However, in general cases, skills under consideration may have such a hierarchical order as the above, for example, for skills S_1, S_2, S_3, S_4, there may be a case with skill patterns $(0, 0, 0, 0)$,$(1, 0, 0, 0)$, $(1, 1, 0, 0)$,$(1, 1, 0, 0)$, $(1, 0, 1, 0)$, $(1, 1, 1, 0)$, $(1, 1, 1, 1)$. For treating such cases, extensions of the latent distance model were proposed by several authors [3–5, 8, 11].

In this chapter, latent class analysis with latent binary variables is discussed. Section 4.2 reviews latent class models for dealing with scale patterns of the latent variables. In Sect. 4.3, the ML estimation procedure for a structured latent class model for explaining learning structures is discussed. Section 4.4 provides numerical examples to demonstrate the analysis. In Sect. 4.5, an approach to consider learning or developmental processes is given. Sections 4.6 and 4.7 consider a method for evaluating mixed ratios of learning processes in a population. In Sect. 4.8, a path analysis in learning and/or developmental structures is treated, and in Sect. 4.9, a numerical example is provided to demonstrate the analysis. Finally, in Sect. 4.10, a summary of the present chapter and discussions on the latent class analysis with binary latent variables are given for leading to further studies to develop the present approaches in the future.

4.2 Latent Class Model for Scaling Skill Acquisition Patterns

In (4.1) and (4.2), let Ω be the sample space of latent variable (skill or trait acquisition) vector $\boldsymbol{S} = (S_1, S_2, \ldots, S_I)$ and let $v(\boldsymbol{s})$ be the latent class proportions with latent variable vector $\boldsymbol{S} = \boldsymbol{s} \in \Omega$, where $\boldsymbol{s} = (s_1, s_2, \ldots, s_I)$. Then, an extended version of latent distance model (3.6) was made as follows [5]:

$$P(\boldsymbol{X} = \boldsymbol{x}) = \sum_{\Omega \ni \boldsymbol{s}} v(\boldsymbol{s}) P(\boldsymbol{X} = \boldsymbol{x}|\boldsymbol{S} = \boldsymbol{s}), \tag{4.3}$$

where

$$P(\boldsymbol{X}=\boldsymbol{x}|\boldsymbol{S}=\boldsymbol{s})=\prod_{i=1}^{I}P(X_i=x_i|S_i=s_i)$$
$$=\prod_{i=1}^{I}\left(\frac{\exp(\alpha_i+s_i\exp(\beta_i))}{1+\exp(\alpha_i+s_i\exp(\beta_i))}\right)^{x_i}\left(\frac{1}{1+\exp(\alpha_i+s_i\exp(\beta_i))}\right)^{1-x_i}$$
$$=\prod_{i=1}^{I}\frac{\exp\{x_i(\alpha_i+s_i\exp(\beta_i))\}}{1+\exp(\alpha_i+s_i\exp(\beta_i))}. \tag{4.4}$$

In which follows, the term "skill" is employed for convenience of the discussion. In the above model, the intrusion (guessing) and omission (forgetting) error probabilities for responding to items i, $P(X_i=1|S_i=0)$ and $P(X_i=0|S_i=1)$, are, respectively, expressed as follows:

$$P(X_i=1|S_i=0)=\frac{\exp(\alpha_i)}{1+\exp(\alpha_i)},\ P(X_i=0|S_i=1)$$
$$=\frac{1}{1+\exp(\alpha_i+\exp(\beta_i))},\ i=1,2,\ldots,I. \tag{4.5}$$

Considering responses to test items, the following inequalities are needed.

$$P(X_i=1|S_i=0)<P(X_i=1|S_i=1),\ i=1,2,\ldots,I. \tag{4.6}$$

The above inequalities are satisfied by the structured model (4.5), so this model is an extension of the following three models.

As reviewed in Chap. 3, in Proctor [11], the intrusion and omission error probabilities were given by

$$P(X_i=1|S_i=0)=P(X_i=0|S_i=1)=\pi_L,\ i=1,2,\ldots,I. \tag{4.7}$$

In this model, the intrusion and omission error probabilities are constant through test items. Following the above model, in Macready and Dayton [3], the following error probabilities are used:

$$P(X_i=1|S_i=0)=\pi_L,\ P(X_i=0|S_i=1)=1-\pi_H,\ i=1,2,\ldots,I. \tag{4.8}$$

In the above model, the intrusion and omission error probabilities are, respectively, constant through the items. In Dayton and Macready [4],

$$P(X_i=1|S_i=0)=P(X_i=0|S_i=1)=\pi_{Li},\ i=1,2,\ldots,I. \tag{4.9}$$

In this model, the intrusion and omission error probabilities are equal for each test item. The above three models do not satisfy the inequalities (4.7) in the parameter estimation, without making any structures as model (4.5). In the next section, an ML estimation procedure for model (4.3) with (4.4) is given according to the EM algorithm.

4.3 ML Estimation Procedure for Model (4.3) with (4.4)

For a practical convenience, it is assumed the sample space of latent variable vector $\boldsymbol{S} = (S_1, S_2, \ldots, S_I)^T$, Ω, includes all skill acquisition patterns, $(0, 0, \ldots, 0), (1, 0, \ldots, 0), \ldots, (1, 1, \ldots, 1)$. Let $\boldsymbol{\phi} = ((v(\boldsymbol{s})), (\alpha_i), (\beta_i))$ be the parameter vector of the latent class model (4.5), the EM algorithm for obtaining the ML estimates of the parameters is given as follows:

EM algorithm

(i) E-step

Let ${}^t\boldsymbol{\phi} = \left(\left({}^t v(\boldsymbol{s})\right), \left({}^t\alpha_i\right), \left({}^t\beta_i\right)\right)^T$ be the estimate of parameter vector $\boldsymbol{\phi}$ at the t th iteration in the EM algorithm. Then, the conditional expectations of complete data $(n(\boldsymbol{x}, \boldsymbol{s}))$ for given parameters ${}^t\boldsymbol{\phi} = \left(\left({}^t v(\boldsymbol{s})\right), \left({}^t\alpha_i\right), \left({}^t\beta_i\right)\right)$ are calculated in the $(t+1)$ th iteration as follows:

$$ {}^{t+1}n(\boldsymbol{x}, \boldsymbol{s}) = n(\boldsymbol{x}) \frac{{}^t v(\boldsymbol{s}) \prod_{i=1}^{I} {}^t P(\boldsymbol{X} = \boldsymbol{x} | \boldsymbol{S} = \boldsymbol{s})}{\sum_{\boldsymbol{u}} {}^t v(\boldsymbol{u}) \prod_{i=1}^{I} {}^t P(\boldsymbol{X} = \boldsymbol{x} | \boldsymbol{S} = \boldsymbol{u})}, \ \boldsymbol{s} \in \Omega, \tag{4.10} $$

where

$$ {}^t P(\boldsymbol{X} = \boldsymbol{x} | \boldsymbol{S} = \boldsymbol{s}) = \frac{\exp\left\{x_i\left({}^t\alpha_i + s_i \exp\left({}^t\beta_i\right)\right)\right\}}{1 + \exp({}^t\alpha_i + s_i \exp({}^t\beta_i))}, \ x_i = 0, 1. $$

In (4.10), notation $\sum_{\boldsymbol{u}}$ implies the summation over all patterns $\boldsymbol{u} \in \Omega$.

(ii) M-step

The loglikelihood function of parameter vector $\boldsymbol{\phi}$ for complete data $\left({}^{t+1}n(\boldsymbol{x}, \boldsymbol{s})\right)$ is given by

$$ \begin{aligned} Q\left(\boldsymbol{\phi} | {}^t\boldsymbol{\phi}\right) &= l\left(\boldsymbol{\phi} | \left({}^{t+1}n(\boldsymbol{x}, \boldsymbol{s})\right)\right) = \sum_{\boldsymbol{s}} \sum_{\boldsymbol{x}} {}^{t+1}n(\boldsymbol{x}, \boldsymbol{s}) \log v(\boldsymbol{s}) \\ &+ \sum_{\boldsymbol{s}} \sum_{\boldsymbol{x}} {}^{t+1}n(\boldsymbol{x}, \boldsymbol{s}) \left[\sum_{i=1}^{I} \{x_i(\alpha_i + s_i \exp(\beta_i)) - \log(1 + \exp(\alpha_i + s_i \exp(\beta_i)))\} \right]. \end{aligned} \tag{4.11} $$

Based on a similar discussion in the previous chapter, we have the estimates ${}^{t+1}v(\boldsymbol{s})$ as follows:

$$ {}^{t+1}v(\boldsymbol{s}) = \frac{\sum_{\boldsymbol{x}} {}^{t+1}n(\boldsymbol{x}, \boldsymbol{s})}{N}, \boldsymbol{s} \in \Omega. \tag{4.12} $$

With respect to parameters α_i and β_i, the Newton–Raphson method has to be employed for maximizing $Q(\boldsymbol{\phi}|{}^t\boldsymbol{\phi})$ in the M-step. Let $\boldsymbol{\phi}_{(\alpha,\beta)} = ((\alpha_i), (\beta_i))$, and let u th iterative value of $\boldsymbol{\phi}_{(\alpha,\beta)}$ in the M-step be $\boldsymbol{\phi}_{(\alpha,\beta)u} = ((\alpha_{iu}), (\beta_{iu}))$, where $\boldsymbol{\phi}_{(\alpha,\beta)1} = \left(\left({}^t\alpha_i\right), \left({}^t\beta_i\right)\right)$, Then, $\boldsymbol{\phi}_{(\alpha,\beta)u+1}$ is obtained as follows:

$$ \boldsymbol{\phi}_{(\alpha,\beta)u+1} = \boldsymbol{\phi}_{(\alpha,\beta)u} - \boldsymbol{H}_u^{-1}\boldsymbol{g}_u, u = 1, 2, \ldots, \tag{4.13} $$

where $\boldsymbol{g}_u$ and $\boldsymbol{H}_u$ are values of the gradient vector and the Hessian matrix at $\boldsymbol{\phi} = \left(\left({}^{t+1}v(\boldsymbol{s})\right), \boldsymbol{\phi}_{(\alpha,\beta)u}\right)$. From this algorithm, we can get $u \to \infty \lim \boldsymbol{\phi}_{(\alpha,\beta)u} = \left(\left({}^{t+1}\alpha_i\right), \left({}^{t+1}\beta_i\right)\right)$.

Remark 4.1 The gradient vector and the Hessian matrix in the above M-step are set as follows:

$$ \boldsymbol{g} = \begin{pmatrix} \left(\frac{\partial Q(\boldsymbol{\phi}|{}^t\boldsymbol{\phi})}{\partial\alpha_i}\right) \\ \left(\frac{\partial Q(\boldsymbol{\phi}|{}^t\boldsymbol{\phi})}{\partial\alpha_i}\right) \end{pmatrix}, \boldsymbol{H} = \begin{pmatrix} \left(\frac{\partial^2 Q(\boldsymbol{\phi}|{}^t\boldsymbol{\phi})}{\partial\alpha_i\partial\alpha_j}\right) & \left(\frac{\partial^2 Q(\boldsymbol{\phi}|{}^t\boldsymbol{\phi})}{\partial\alpha_i\partial\beta_i}\right) \\ \left(\frac{\partial^2 Q(\boldsymbol{\phi}|{}^t\boldsymbol{\phi})}{\partial\beta_i\partial\alpha_i}\right) & \left(\frac{\partial^2 Q(\boldsymbol{\phi}|{}^t\boldsymbol{\phi})}{\partial\beta_i\partial\beta_j}\right) \end{pmatrix}, \tag{4.14} $$

where

$$ \frac{\partial Q(\boldsymbol{\phi}|{}^t\boldsymbol{\phi})}{\partial\alpha_i} = \sum_{\boldsymbol{s}}\sum_{\boldsymbol{x}} {}^{t+1}n(\boldsymbol{x}, \boldsymbol{s})(x_i - P(X_i = 1|S_i = s_i)), $$
$$ i = 1, 2, \ldots, I; $$

$$ \frac{\partial Q(\boldsymbol{\phi}|{}^t\boldsymbol{\phi})}{\partial\beta_i} = \sum_{\boldsymbol{s}}\sum_{\boldsymbol{x}} {}^{t+1}n(\boldsymbol{x}, \boldsymbol{s})(x_i - P(X_i = 1|S_i = s_i))s_i\exp(\beta_i), $$
$$ i = 1, 2, \ldots, I; $$

$$ \frac{\partial^2 Q(\boldsymbol{\phi}|{}^t\boldsymbol{\phi})}{\partial a_i^2} = -\sum_{\boldsymbol{s}}\sum_{\boldsymbol{x}} {}^{t+1}n(\boldsymbol{x}, \boldsymbol{s})P(X_i = 1|S_i = s_i)(1 - P(X_i = 1|S_i = s_i)), $$
$$ i = 1, 2, \ldots, I; $$

$$ \frac{\partial^2 Q(\boldsymbol{\phi}|{}^t\boldsymbol{\phi})}{\partial\alpha_i\partial\beta_i} = -\sum_{\boldsymbol{s}}\sum_{\boldsymbol{x}} {}^{t+1}n(\boldsymbol{x}, \boldsymbol{s})P(X_i = 1|S_i = s_i)(1 - P(X_i = 1|S_i = s_i))s_i\exp(\beta_i), $$
$$ i = 1, 2, \ldots, I; $$

$$\frac{\partial^2 Q(\phi|^t\phi)}{\partial \beta_i^2} = \sum_{s} \sum_{x}^{t+1} n(x, s)\{x_i - P(X_i = 1|S_i = s_i)$$
$$- P(X_i = 1|S_i = s_i)(1 - P(X_i = 1|S_i = s_i))s_i \exp(\beta_i)\}s_i \exp(\beta_i),$$
$$i = 1, 2, \ldots, I;$$

$$\frac{\partial^2 Q(\boldsymbol{\phi}|^s\boldsymbol{\phi})}{\partial \alpha_i \partial \alpha_j} = \frac{\partial^2 Q(\boldsymbol{\phi}|^s\boldsymbol{\phi})}{\partial \alpha_i \partial \beta_j} = \frac{\partial^2 Q(\boldsymbol{\phi}|^s\boldsymbol{\phi})}{\partial \beta_i \partial \beta_j} = 0, i \neq j.$$

The above algorithm has been made, where the latent sample space Ω has all the 2^I skill acquisition patters; however, to identify the latent class model, the number of latent classes A are restricted by

$$A < 2^I - 2I. \tag{4.15}$$

The above algorithm has the following property. If we set the initial value of class proportion $v(s)$ as ${}^0v(s) = 0$, from (4.10) we have ${}^1n(x, s) = 0$ for all the manifest response patterns x. From (4.12) it follows that ${}^1v(s) = 0$, and inductively, we obtain

$${}^tv(s) = 0, t = 1.2, 3, \ldots$$

Hence, if we set ${}^0v(s) = 0$ for all skill acquisition patterns $s \in \Omega - \Omega_0$ in order to identify the model, where Ω_0 is a set of skill acquisition patterns assumed beforehand, the class proportions are automatically set as zeroes, and the above algorithm can work effectively to get the ML estimates of the model parameters. For example, if for $I = 4$, we set

$$\Omega_0 = \{(0, 0, 0, 0), (1, 0, 0, 0), (1, 1, 0, 0), (1, 1, 1, 0), (1, 1, 1, 1)\},$$

the above algorithm can be used for estimating the latent distance model. In the present latent class analysis, it is meaningful to detect latent classes (skill acquisition patterns) s with positive class proportions $v(s) > 0$. In the next section, through numerical examples with practical data sets in Chap. 2, an exploratory method for determining the latent classes is demonstrated.

4.4 Numerical Examples (Exploratory Analysis)

By using the Stouffer-Toby data (Table 2.1), McHugh data (Table 2.5), and Lazarsfeld-Stouffer data (Table 2.7), the present latent class analysis is demonstrated. From restriction (4.15) with $I = 4$, we have $A < 8$, so the maximum number of latent classes is seven. From this, considering response data in Tables 2.1, , 2.5, and 2.7, the following skill acquisition patterns (latent classes) are assumed in the data sets

Table 4.1 The sets of initial skill acquisition patterns (Ω_0) for the three data sets

Stouffer-Toby data	(0, 0, 0, 0), (0, 1, 1, 0), (0, 0, 0, 1), (0, 1, 0, 1), (1, 1, 0, 1), (0, 0, 1, 1), (1, 1, 1, 1)
McHugh data	(0, 0, 0, 0), (0, 1, 0, 0), (1, 1, 0, 0), (1, 1, 1, 0), (0, 0, 1, 1), (0, 1, 1, 1), (1, 1, 1, 1)
Lazarsfeld-Stouffer data	(0, 0, 0, 0), (0, 0, 1, 0), (0, 0, 0, 1), (0, 1, 0, 1), (0, 0, 1, 1), (0, 1, 1, 1), (1, 1, 1, 1)

(Table 4.1) as the initial skill acquisition patterns (latent classes). In order to select the best model, a backward elimination procedure is used. Let Ω_0 be the initial set of skill acquisition patterns, for example, in Table 4.1; $M(\Omega_0)$ be the initial model with Ω_0; and let $\widehat{M}(\Omega_0)$ be the ML estimate of $M(\Omega_0)$. According to $\widehat{M}(\Omega_0)$, the latent class with the minimum proportion $\hat{v}(s)$ is deleted from the initial skill acquisition patterns. Let Ω_1 be the set of the patterns left, and let $M(\Omega_1)$ be the model with Ω_1. Then, the ML estimates $\widehat{M}(\Omega_0)$ and $\widehat{M}(\Omega_1)$ are compared with the log likelihood ratio test or the Pearson chi-square test, and if the results are statistically significant with significance level α, then, $M(\Omega_1)$ is accepted and the procedure continues similarly by setting Ω_1 as the initial skill acquisition pattern set, whereas if the results are not significant, the procedure stops and model $M(\Omega_0)$ is selected as a most suitable model. The algorithm is shown as follows:

Backward Elimination Procedure

(i) Set Ω_0 as the initial skill acquisition patterns.
(ii) Obtain the ML estimate $\widehat{M}(\Omega_0)$.
(iii) Delete the pattern $\boldsymbol{s}_k$ with the minimum value of $\hat{v}(\boldsymbol{s})$ from Ω_k and set $\Omega_{k+1} = \Omega_k \setminus \{\boldsymbol{s}_k\}$
(iv) Obtain $\widehat{M}(\Omega_{k+1})$
(v) If $\widehat{M}(\Omega_{k+1})$ is accepted for (better than) $\widehat{M}(\Omega_k)$ according to the loglikelihood ratio test for the relative goodness-of-fit to data set, go to (iii) by substituting k for $k+1$, if not, the procedure stops.

According to the above procedure, we have the final models shown in Table 4.2. For Stouffer-Toby data and Lazarsfel-Stouffer data, the results are the same as those in Tables 2.3 and 2.9, respectively. It may be said that concerning Stouffer-Toby data, there exist latent "universalistic" and "particularistic" states for responding to the test items, and with respect to Lazarsfeld-Stouffer data, latent "favorable" and "unfavorable" states to the Army. Hence, it means that all four skills (traits) are equal, i.e., $S_1 = S_2 = S_3 = S_4$. The learning structure is expressed as.

$$(0, 0, 0, 0) \rightarrow (1, 1, 1, 1).$$

For McHugh data, the results are interpreted as $S_1 = S_2$ and $S_3 = S_4$, and the learning structure can be expected as in Fig. 4.1, and the following two learning processes can be assumed:

$$\text{(i) } (0, 0, 0, 0) \rightarrow (1, 1, 0, 1) \rightarrow (1, 1, 1, 1) \text{ and}$$

Table 4.2 The results of the analysis of the three data sets

			Item positive response probability			
	Pattern*	Proportion**	X_1	X_2	X_3	X_4
Stouffer-Toby	(0, 0, 0, 0)	0.279	0.007	0.060	0.073	0.231
	(1, 1, 1, 1)	0.721	0.286	0.670	0.646	0.868
	Test of GF***	$G^2 = 2.720, df = 6, P = 0.843$				
McHugh	(0, 0, 0, 0)	0.396	0.239	0.244	0.112	0.204
	(1, 1, 0, 0)	0.077	0.894	0.996	0.112	0.204
	(0, 0, 1, 1)	0.200	0.239	0.244	0.979	0.827
	(1, 1, 1, 1)	0.327	0.894	0.996	0.979	0.827
	Test of GF***	$G^2 = 1.100, df = 4, P = 0.894$				
Lazarsfeld-Stouffer	(0, 0, 0, 0)	0.445	0.093	0.386	0.442	0.499
	(1, 1, 1, 1)	0.555	0.572	0.818	0.906	0.944
	Test of GF***	$G^2 = 8.523, df = 6, P = 0.202$				

* Skill Acquisition Pattern; ** Class Proportion; ***Test of Goodness-of-Fit

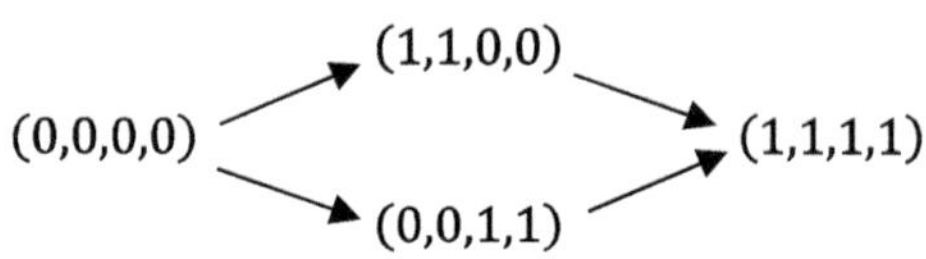

Fig. 4.1 The learning structure in McHugh data set

$$\text{(ii)}\ (0, 0, 0, 0) \rightarrow (0, 0, 1, 1) \rightarrow (1, 1, 1, 1) \tag{4.16}$$

In this case, it is significant to discuss the proportions of subpopulations according to the above two learning processes. The topic is treated in the next section.

4.5 Dynamic Interpretation of Learning (Skill Acquisition) Structures

Let S_i be skill acquisition states of skill $i = 1, 2, \ldots, I$, and let skill i be prerequisite to skill $i + 1, i = 1, 2, \ldots, I - 1$. To discuss a dynamic interpretation of learning structures, the following notation is introduced:

$$S_1 \rightarrow S_2 \rightarrow \cdots \rightarrow S_I. \tag{4.17}$$

In the above prerequisite relation, the sample space of $\boldsymbol{S} = (S_1, S_2, \ldots, S_I)$ is

$$\Omega = \{(0, 0, \ldots, 0), (1, 0, \ldots, 0), \ldots, (1, 1, \ldots, 1)\}, \tag{4.18}$$

and the space is called a learning space in this book. As in the previous section, notation (sequence) (4.17) can also be expressed by using skill acquisition patterns in learning space Ω:

$$(0, 0, \ldots, 0) \to (1, 0, \ldots, 0) \to \cdots \to (1, 1, \ldots, 1).$$

In this case, the model is the latent distance model discussed in the previous chapter. Hence, the conditional probabilities $P(S_{i+1} = s_{i+1}|(S_1, S_2, \ldots, S_i) = (s_1, s_2, \ldots, s_i))$ are given by

$$P(S_{i+1} = s_{i+1}|(S_1, S_2, \ldots, S_i) = (s_1, s_2, \ldots, s_i)) = \begin{cases} 0 & (s_i = 0, s_{i+1} = 1) \\ 1 & (s_i = 0, s_{i+1} = 0) \\ P(S_{i+1} = s_{i+1}|S_i = s_i) & (s_i = 1) \end{cases}, i = 1, 2, \ldots, i.$$

Since in the sequence of latent variables $\{S_i\}$ (4.17), S_{i+1} depends on only the state of S_i from the above discussion, we have the following theorem:

Theorem 4.1 Sequence (4.17) is a Markov chain. ■

From the above discussion, prerequisite relations among skills to be scaled can be interpreted as learning processes shown in (4.17). Below, "structure" and "process" will be used as the same in appropriate cases. Let

$$q_{s_i s_{i+1}, i} = P(S_{i+1} = s_{i+1}|S_i = s_i), i = 1, 2, \ldots, I - 1.$$

The transition matrix $\boldsymbol{Q}_i$ from S_i to S_{i+1} is given by

$$\boldsymbol{Q}_i = \begin{pmatrix} 1 & 0 \\ q_{10,i} & q_{11,i} \end{pmatrix}, i = 1, 2, \ldots, I - 1. \tag{4.19}$$

Although sequence $\{S_i\}$ are not observed sequentially at points in time, Theorem 4.1 induces a dynamic interpretation of learning space (4.18). The transition probabilities $q_{11,i}$ imply the intensities of skill acquisition of skill $i + 1$ given skill i, $i = 1, 2, \ldots, I - 1$. We can give it the following dynamic interpretation: if an individual acquires skill i, then, the individual acquires skill $i + 1$ with probability $q_{11,i}$. We have the following theorem [6]:

Theorem 4.2 The transition probabilities $q_{10,i}$ and $q_{11,i}$ of Markov chain (4.17) are expressed as follows:

$$q_{10,i} = \frac{v(\overbrace{1, 1, \ldots, 1}^{i}, 0, \ldots, 0)}{\sum_{k=i}^{I} v(\underbrace{1, 1, \ldots, 1}_{k}, 0, \ldots, 0)}, q_{11,i} = 1 - q_{10,i}, i = 1, 2, \ldots, 1 - 1.$$

Proof $q_{10,i} = P(S_{i+1} = 0|S_i = 1) = \frac{P(S_i=1,S_{i+1}=0)}{P(S_i=1)}$.

Let subset $\Omega_i \subset \Omega$ be defined by

$$\Omega_i = \left\{ (\overbrace{1, 1, \ldots, 1}^{k}, 0, \ldots, 0), k = i, i+1, \ldots, I \right\}.$$

Then, in Markov chain (4.17), it follows that

$$S_i = 1 \iff \boldsymbol{S} = (S_1, S_2, \ldots, S_I) \in \Omega_i$$

and

$$S_i = 1, S_{i+1} = 0 \Leftrightarrow \boldsymbol{S} = (\overbrace{1, 1, \ldots, 1}^{i}, 0, \ldots, 0).$$

From this, we have

$$P(S_i = 1) = P((S_1, S_2, \ldots, S_I) \in \Omega_i) = \sum_{k=i}^{I} v(\underbrace{1, 1, \ldots, 1}_{k}, 0, \ldots, 0),$$

$$P(S_i = 1, S_{i+1} = 0) = P\left((S_1, S_2, \ldots, S_I) = (\underbrace{1, 1, \ldots, 1}_{i}, 0, \ldots, 0) \right)$$

$$= v\left((\underbrace{1, 1, \ldots, 1}_{i}, 0, \ldots, 0) \right).$$

Thus, the theorem follows. ■

The probabilities $q_{11,i}$ are regarded as the path coefficients relating to path $S_i \to S_{i+1}, i = 1, 2, \ldots, I-1$, and the direct effects of S_i on S_{i+1} are defined by the path coefficients. The path diagram (4.17) can be illustrated as follows:

$$S_1 \overset{q_{11,1}}{\to} S_2 \overset{q_{11,2}}{\to} \cdots \overset{q_{11,I-1}}{\to} S_I.$$

Moreover, we have the following theorem.

Theorem 4.3 In (4.17), the following formulae hold true:

$$P(S_j = 1|S_i = 1) = \prod_{k=i}^{j-1} q_{11,k}, j > i. \tag{4.20}$$

Proof Since sequence (4.17) is a Markov chain with transition matrices (4.19), the theorem follows. ■

The probabilities in (4.20) are calculated by multiplying the related path coefficients, so we have the following definition:

Definition 4.1 In learning structure (4.17), for $j > i$, probabilities $P(S_j = 1|S_i = 1)$ in (4.20) are defined as the pathway effects of S_i on S_j through path

$$S_i \overset{q_{11,i}}{\rightarrow} S_{i+1} \overset{q_{11,i+1}}{\rightarrow} \cdots \overset{q_{11,j-1}}{\rightarrow} S_j.$$

The pathway effects are denoted by $e_{path}(S_i \rightarrow S_{i+1} \rightarrow \cdots \rightarrow S_j)$.

In the above definition, paths $S_i \rightarrow S_{i+1} \rightarrow \cdots \rightarrow S_j$ are partial paths of (4.17). If path $S_{i_1} \rightarrow S_{i_2} \rightarrow \cdots \rightarrow S_{i_k}$ is not a partial path of (4.17), then, we set

$$e_{path}(S_{i_1} \rightarrow S_{i_2} \rightarrow \cdots \rightarrow S_{i_k}) = 0.$$

The above discussion is applied to the results of the latent distance analysis (Table 3.2) of Stouffer-Toby data set (Table 2.1). Since

$$\hat{v}(0,0,0,0) = 0.296, \hat{v}(1,0,0,0) = 0.344, \hat{v}(1,1,0,0) = 0.103,$$

$$\hat{v}(1,1,1,0) = 0.049, \hat{v}(1,1,1,1) = 0.208,$$

from Theorem 4.2, we have

$$\hat{q}_{10,1} = \frac{\hat{v}(1,0,0,0)}{\hat{v}(1,0,0,0) + \hat{v}(1,1,0,0) + \hat{v}(1,1,1,0) + \hat{v}(1,1,1,1)} = 0.489,$$
$$\hat{q}_{11,1} = 0.511,$$

$$\hat{q}_{10,2} = \frac{\hat{v}(1,1,0,0)}{\hat{v}(1,1,0,0) + \hat{v}(1,1,1,0) + \hat{v}(1,1,1,1)} = 0.286, \hat{q}_{11,2} = 0.714,$$

$$\hat{q}_{10,3} = \frac{\hat{v}(1,1,1,0)}{\hat{v}(1,1,1,0) + \hat{v}(1,1,1,1)} = 0.191, \hat{q}_{11,3} = 0.809.$$

The path coefficients are illustrated with the sequence of $S_i, i = 1,2,3,4$, and we have

$$S_1 \overset{0.511}{\rightarrow} S_2 \overset{0.714}{\rightarrow} S_3 \overset{0.809}{\rightarrow} S_4. \tag{4.21}$$

According to Theorem 4.3, for example, the pathway effect of S_1 on S_4 is calculated as

Table 4.3 Pathway effects of S_i on S_j, $i < j$ in sequence (4.21)

	S_2	S_3	S_4
S_1	0.511	0.375	0.295
S_2	–	0.714	0.578
S_3	–	–	0.809

$$0.511 \times 0.714 \times 0.809 = 0.295.$$

All the pathway effects in sequence (4.21) are shown in Table 4.3.

For McHugh data, the results of latent class analysis show there are two learning processes (4.16) (Fig. 4.1). In this case, for $S_1(=S_2)$ and $S_3(=S_4)$, the learning structure is a mixture of the following processes:

$$(i) S_1 \to S_3, (ii) S_3 \to S_1, \tag{4.22}$$

and it is meaningful to consider the mixed ratios of the learning processes in the population. To treat such cases, the next section discusses general learning structures.

4.6 Estimation of Mixed Proportions of Learning Processes

Suppose that there exist the following three learning processes:

$$\begin{cases} (i) S_1 \to S_2 \to S_3 \to S_4, \\ (ii) S_3 \to S_2 \to S_1 \to S_4, \\ (iii) S_3 \to S_1 \to S_2 \to S_4. \end{cases} \tag{4.23}$$

Then, the sample space of (S_1, S_2, S_3, S_4) is

$$\begin{aligned} \Omega = & \{(0,0,0,0), (1,0,0,0), (0,0,1,0), (1,1,0,0), \\ & (1,0,1,0), (0,1,1,0)(1,1,1,0), (1,1,1,1)\}, \end{aligned} \tag{4.24}$$

so the above learning structure is expressed with skill acquisition patterns (s_1, s_2, s_3, s_4) (Fig. 4.2). It is assumed that a population is divided into three subpopulations, each of which depends on one of the three learning processes (structure) (4.23). Let $v(s_1, s_2, s_3, s_4, Process\ k)$ be the proportions of individuals with skill acquisition patters (s_1, s_2, s_3, s_4) and learning process k, $k = 1, 2, 3$. Then, in general, it follows that

$$v(s_1, s_2, s_3, s_4) = \sum_{k=1}^{3} v(s_1, s_2, s_3, s_4, Process\ k). \tag{4.25}$$

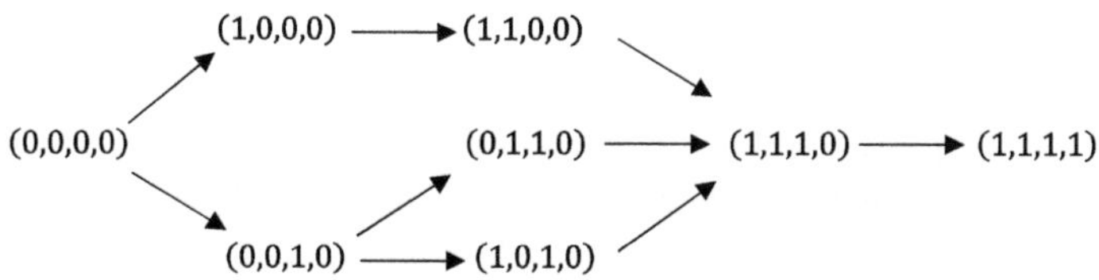

Fig. 4.2 Path diagram of (4.23) based on the sample space (4.24)

In Fig. 4.2, the following equations hold:

$$\begin{cases} v(0,0,0,0) = \sum_{k=1}^{3} v(0,0,0,0, Process\ k) \\ v(1,1,1,0) = \sum_{k=1}^{3} v(1,1,1,0, Process\ k), \\ v(1,1,1,1) = \sum_{k=1}^{3} v(1,1,1,1, Process\ k) \end{cases} \quad (4.26)$$

$$v(0,0,1,0) = v(0,0,1,0, Process\ 2) + v(0,0,1,0, Process\ 3), \quad (4.27)$$

$$\begin{cases} v(1,0,0,0) = v(1,0,0,0, Process\ 1) \\ v(1,1,0,0) = v(1,1,0,0, Process\ 1) \\ v(0,1,1,0) = v(0,1,1,0, Process\ 2) \\ v(1,0,1,0) = v(1,0,1,0, Process\ 3) \end{cases}. \quad (4.28)$$

In (4.23), although each sequence is a Markov chain, parameters $v(s_1, s_2, s_3, s_4, Process\ k)$ in (4.26) and (4.27) are not identified, so we have to impose a constraint on the parameters. Let $w_k, k = 1, 2, 3$ be proportions of subpopulations with learning processes (4.23). Then,

$$\sum_{k=1}^{3} w_k = 1.$$

In order to identify the parameters $v(s_1, s_2, s_3, s_4, Process\ k)$, in this chapter, the following constraint is placed on the parameters. If a skill acquisition pattern is derived from some learning processes, it is assumed that the proportions of individuals with the skill acquisition patterns derived from the learning processes are in proportion to the related proportions w_k. For example, skill acquisition pattern $(0, 0, 1, 0)$ comes from learning processes 2 and 3 (4.27), so we have the following equations:

$$\begin{cases} v(0,0,1,0, Process\ 2) = \frac{w_2}{w_2+w_3} v(0,0,1,0) \\ v(0,0,1,0, Process\ 3) = \frac{w_3}{w_2+w_3} v(0,0,1,0) = v(0,0,1,0) - v(0,0,1,0, Process\ 2) \end{cases}.$$

Under the above assumption, for learning process 1, we obtain

$$\begin{cases} v(0,0,0,0, Process\ 1) = w_1 v(0,0,0,0) \\ v(1,0,0,0, Process\ 1) = v(1,0,0,0) \\ v(1,1,0,0, Process\ 1) = v(1,1,0,0) \\ v(1,1,1,0, Process\ 1) = w_1 v(1,1,1,0) \\ v(1,1,1,1, Process\ 1) = w_1 v(1,1,1,1) \end{cases}. \tag{4.29}$$

From (4.29), we have

$$\begin{aligned} w_1 &= v(0,0,0,0, Process\ 1) + v(1,0,0,0, Process\ 1) \\ &\quad 7 + v(1,1,0,0, Process\ 1) + v(1,1,1,0, Process\ 1) + v(1,1,1,1, Process\ 1) \\ &= w_1(v(0,0,0,0) + v(1,1,1,0) + v(1,1,1,1)) + v(1,0,0,0) + v(1,1,0,0). \end{aligned} \tag{4.30}$$

Similarly, it follows that

$$\begin{aligned} w_2 &= w_2(v(0,0,0,0) + v(1,1,1,0) + v(1,1,1,1)) \\ &\quad + \frac{w_2}{w_2 + w_3} v(0,0,1,0) + v(0,1,1,0), \end{aligned} \tag{4.31}$$

$$\begin{aligned} w_3 &= w_3(v(0,0,0,0) + v(1,1,1,0) + v(1,1,1,1)) \\ &\quad + \frac{w_3}{w_2 + w_3} v(0,0,1,0) + v(1,0,1,0). \end{aligned} \tag{4.32}$$

In this chapter, the above Eqs. (4.30)–(4.32) are called separating equations for evaluating the mixed proportions $w_k, k = 1, 2, 3$. From the above equations, we get

$$\begin{cases} w_1 = \frac{v(1,0,0,0)+v(1,1,0,0)}{1-(v(0,0,0,0)+v(1,1,1,0)+v(1,1,1,1))} \\ w_2 = (1 - w_1)\frac{v(0,1,1,0)}{v(0,1,1,0)+v(1,0,1,0)} \\ w_3 = 1 - w_1 - w_2 \end{cases}. \tag{4.33}$$

Remark 4.2 Let us consider the solution of separating equations in (4.33). It is seen that

$$0 < w_1 < 1,$$

and in

$$0 < w_2 < 1 - w_1 < 1,$$

so we have

$$0 < w_3 < 1.$$

Hence, solution (4.33) is proper, and such solutions are called proper solutions. Properties of the separating equations are discussed generally in the next section.

By using the above method, the mixed proportions of learning processes 1 and 2 (4.22) in McHugh data are calculated. From Table 4.2, we have the following equations:

$$\begin{cases} w_1 = w_1(v(0,0,0,0) + (1,1,1,1)) + v(1,1,0,0), \\ w_2 = 1 - w_1. \end{cases}$$

From the above equations, we have the following solution:

$$w_1 = \frac{v(1,1,0,0)}{1-(v(0,0,0,0)+v(1,1,1,1))},\ w_2 = \frac{v(0,0,1,1)}{1-(v(0,0,0,0)+v(1,1,1,1))}. \tag{4.34}$$

Hence, from (4.34) and Table 4.2, the estimates of the mixed proportions are calculated as follows:

$$\widehat{w}_1 = \frac{0.077}{1-(0.396+0.327)} = 0.278,\ \widehat{w}_2 = 1 - \widehat{w}_1 = 0.722.$$

4.7 Solution of the Separating Equations

The separating equations introduced in the previous section for estimating the mixed proportions of learning processes are considered in a framework of learning structures. First, the learning structures are classified into two types.

Definition 4.2 If all learning processes in a population have skill acquisition patterns peculiar to them, the learning structure is called a clear learning structure. If not, the learning structure is referred to as an unclear learning structure.

In the above definition, learning structures (4.16) and (4.23) are clear learning structures, as shown in Figs. 4.1 and 4.2. On the other hand, the following learning structure is an unclear one:

$$\begin{cases} Process1(w_1): S_1 \to S_2 \to S_3 \to S_4, \\ Process2(w_2): S_2 \to S_1 \to S_3 \to S_4, \\ Process3(w_3): S_2 \to S_3 \to S_1 \to S_4. \end{cases} \tag{4.35}$$

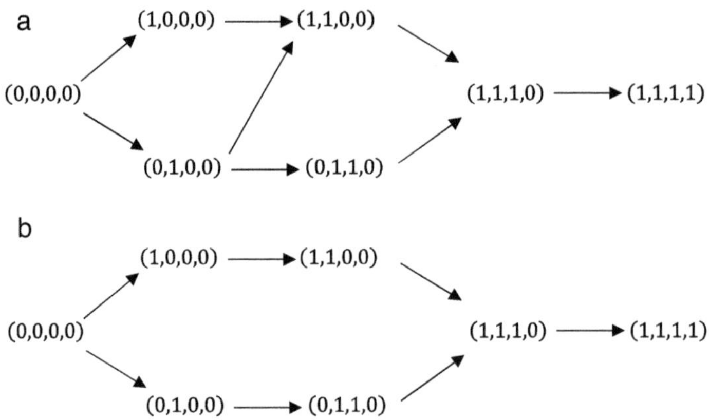

Fig. 4.3 **a** Path diagram of (4.35). **b** Path diagram of the learning structure with Processes 1 and 2 in (4.35)

From the above structure, Fig. 4.3a is made. From the figure, learning processes 1 and 3 have skill acquisition patterns $(1, 0, 0, 0)$ and $(0, 1, 1, 0)$ peculiar to them, respectively; however, there are no skill acquisition patterns peculiar to Process 2. Even if Process 2 is deleted from (4.35), the sample space of (S_1, S_2, S_3, S_4) is the same as (4.35); however, the structure is expressed as in Fig. 4.3b, then, the structure is a clear one. With respect to the learning structure (4.35), from Fig. 4.3a, we have the following separating equations:

$$w_1 = \frac{w_1}{w_1 + w_2} v(1, 1, 0, 0) + w_1 (v(0, 0, 0, 0) + v(1, 1, 1, 0) + v(1, 1, 1, 1)) \\ + v(1, 0, 0, 0),$$

$$w_2 = \frac{w_2}{w_1 + w_2} v(1, 1, 0, 0) + \frac{w_2}{w_2 + w_3} v(0, 0, 1, 0) \\ + w_2 (v(0, 0, 0, 0) + v(1, 1, 1, 0) + v(1, 1, 1, 1)),$$

$$w_3 = \frac{w_3}{w_2 + w_3} v(0, 0, 1, 0) + w_3 (v(0, 0, 0, 0) + v(1, 1, 1, 0) + v(1, 1, 1, 1)) \\ + v(0, 1, 1, 0).$$

From the above equations, we have the following solution:

$$w_1 = \frac{v(1, 0, 0, 0)}{v(1, 0, 0, 0) + v(1, 1, 0, 0)}, w_2 = 1 - w_1 - w_3, w_3 = \frac{v(0, 1, 1, 0)}{v(0, 1, 0, 0) + v(0, 1, 1, 0)}. \quad (4.36)$$

In this solution, we see that $0 < w_1 < 1, 0 < w_3 < 1$, however, there are cases where condition $0 < w_2 < 1$ does not hold, for example, if $v(1, 0, 0, 0) = 0.1, v(1, 1, 0, 0) = 0.1, v(0, 1, 1, 0) = 0.3, v(0, 1, 0, 0) = 0.1$, then, we have

$$w_1 = 0.5, w_2 = -0.25, w_3 = 0.75.$$

The above solution is improper. If we set $w_2 = 0$, that is, a clear learning structure is shown in Fig. 4.3b, the mixed proportions w_k are calculated as follows:

$$w_1 = \frac{v(1,0,0,0) + v(1,1,0,0)}{v(1,0,0,0) + v(0,1,0,0) + v(1,1,0,0) + v(0,1,1,0)}, w_2 = 0, w_3 = 1 - w_1.$$

The above solution is viewed as a proper solution for learning structure with Processes 1 and 3 in (4.35), and is referred to as a boundary solution for learning structure (4.35). With respect to the separating equations, in general we have the following theorem:

Theorem 4.4 Let a clear learning structure be made of K learning processes, and let $w_k, k = 1, 2, \ldots, K$ be the mixed proportions of the processes. Then, the set of separating equations has a proper solution such that

$$w_k > 0, k = 1, 2, \ldots, K,$$

and

$$\sum_{k=1}^{K} w_k = 1. \tag{4.37}$$

Proof Suppose that a clear learning structure consists of Process 1, Process 2,..., and Process K. Let Ω be the sample space of skill acquisition patterns $\boldsymbol{s} = (s_1, s_2, \ldots, s_I)$ in the clear learning structure; let $\Omega_k (\neq \phi), k = 1, 2, \ldots, K$ be the set of all skill acquisition patterns peculiar to Process k, and let w_k be the proportions of individuals according to Process k in the population. Then, the separating equations are expressed as follows:

$$w_k = \sum_{s \in \Omega_k} v(\boldsymbol{s}) + f_k\left(w_1, w_2, \ldots, w_K | v(\boldsymbol{s}), \boldsymbol{s} \in \Omega \setminus \bigcup_{k=1}^{K} \Omega_k\right), k = 1, 2, \ldots, K, \tag{4.38}$$

where $f_k\left(w_1, w_2, \ldots, w_K, s \in \Omega \setminus \bigcup_{k=1}^{K} \Omega_k\right)$ are positive and continuous functions of $w_i, k = 1, 2, \ldots, K$, given $v(\boldsymbol{s}), \boldsymbol{s} \in \Omega \setminus \bigcup_{k=1}^{K} \Omega_k$. From $\Omega_k (\neq \phi)$, we have

$$\sum_{s \in \Omega_k} v(\boldsymbol{s}) > 0, k = 1, 2, \ldots, K.$$

Let us consider the following function, $(w_1, w_2, \ldots, w_K) \rightarrow (u_1, u_2, \ldots, u_K)$:

$$u_k = \sum_{s\in\Omega_k} v(s) + f_k\left(w_1, w_2, \ldots, w_K, v(s), s \in \Omega \setminus \bigcup_{k=1}^{K} \Omega_k\right), k = 1, 2, \ldots, K. \tag{4.39}$$

For $w_k > 0, k = 1, 2, \ldots, K$, from (4.39) we have

$$u_k > \sum_{s\in\Omega_k} v(s) > 0, k = 1, 2, \ldots, K$$

and from the definition of the separating equation, it follows that

$$\sum_{k=1}^{K} u_k = 1.$$

From the above discussion, function $(w_1, w_2, \ldots, w_K) \to (u_1, u_2, \ldots, u_K)$ is continuous on domain

$$D = \left\{(w_1, w_2, \ldots, w_K) | \sum_{k=1}^{K} w_k = 1; w_k \geq \sum_{s\in\Omega_k} v(s) > 0\right\}.$$

The above function can be regarded as function $D \to D$. Since set D is convex and closed, from the Brouwer's fixed point theorem, there exists a point $(w_{01}, w_{02}, \ldots, w_{0K})$ such that

$$w_{0k} = \sum_{s\in\Omega_k} v(s) + f_k\left(w_{01}, w_{02}, \ldots, w_{0K} | v(s), s \in \Omega \setminus \bigcup_{k=1}^{K} \Omega_k\right) > 0, k = 1, 2, \ldots, K.$$

Hence, the theorem follows: ■

In general, we have the following theorem.

Theorem 4.5 Let a learning structure be made of K learning processes, and let $w_k, k = 1, 2, \ldots, K$ be the mixed proportions of the processes. Then, the set of separating equations has a solution such that

$$w_k \geq 0, k = 1, 2, \ldots, K, \tag{4.40}$$

and

$$\sum_{k=1}^{K} w_k = 1.$$

Proof In the clear learning structure, from Theorem 4.4, the theorem follows. On the other hand, in the unclear learning structure, deleting some learning processes from the structure, that is, setting the mixed proportions of the correspondent learning processes as zeroes, $w_k = 0$, then, we have a clear learning structure that has the same sample space of skill acquisition patterns as the original unclear learning structure. Then, we have the solution as in (4.40). This completes the theorem. ■

A general method for obtaining solutions of the separating equations is given. In general, a system of separating equations is expressed as follows:

$$w_k = g_k(w_1, w_2, \ldots, w_K | v(\mathbf{s}), \mathbf{s} \in \Omega), k = 1, 2, \ldots, K. \tag{4.41}$$

The following function $(w_1, w_2, \ldots, w_K) \to (u_1, u_2, \ldots, u_K)$:

$$u_k = g_k(w_1, w_2, \ldots, w_K | v(\mathbf{s}), \mathbf{s} \in \Omega), k = 1, 2, \ldots, K \tag{4.42}$$

is viewed as a continuous function $C \to C$, where C is an appropriate convex and closed set, for example, in learning structure (4.35), from Fig. 3a, we can set

$$C = \left\{ (w_1, w_2, w_3) | \sum_{k=1}^{3} w_k = 1, w_1 + w_2 \geq v(1,0,0,0) + v(1,1,0,0), w_3 \geq v(0,1,1,0) \right\},$$

and the above set is convex and closed. Then, function (4.42) has a fixed point $(w_1, w_2, w_3) \in C$. From this, the fixed point can be obtained as a convergence value of the following sequence $(w_{n1}, w_{n2}, \ldots, w_{nK}), n = 1, 2, \ldots$:

$$w_{n+1,k} = g_k(w_{n1}, w_{n2}, \ldots, w_{nK} | v(\mathbf{s}), \mathbf{s} \in \Omega), k = 1, 2, \ldots, K; n = 1, 2, \ldots. \tag{4.43}$$

4.8 Path Analysis in Learning Structures

Let $S_i, i = 1, 2, \ldots, I$ be skill acquisitions of skill i, and let $S_{k_j}, j = 1, 2, \ldots, I$ be those of skill k_j in Process $k, k = 1, 2, \ldots, K$, where $\{S_i, i = 1, 2, \ldots, I\} = \{S_{k_j}, j = 1, 2, \ldots, I\}$. Then, we have the following learning processes:

$$Process\ k(w_k) : S_{k_1} \to S_{k_2} \to \cdots \to S_{k_I}, k = 1, 2, \ldots, K; \tag{4.44}$$

$$\sum_{k=1}^{K} w_k = 1$$

where w_k are proportions of subpopulations with Processes k. The pathway effects of path $S_i \to S_j$ in Process k are defined according to Definition 4.1, and in general, path coefficients for path $S_i \to S_j$ in (4.44) are defined as follows [6].

Definition 4.3 Let $e_{path}\left(S_i \to S_j | Process\ k\right)$ be the pathway effects in Process $k, k = 1, 2, \ldots, K$. Then, the pathway effects of $S_i \to S_j, i \neq j$ are defined by

$$e_{path}\left(S_i \to S_j\right) = \sum_{k=1}^{K} w_k e_{path}\left(S_i \to S_j | Process\ k\right). \tag{4.45}$$

The effects are the probabilities that paths $S_i \to S_j$ exist in the population (learning structure). By using the above definition, path coefficients in a general learning structure (4.44) can be calculated. In (4.35), for example, since

$$e_{path}(S_1 \to S_2 | Process\ 2) = e_{path}(S_1 \to S_2 | Process\ 3) = 0,$$

we have

$$e_{path}(S_1 \to S_2) = w_1 e_{path}(S_1 \to S_2 | Process\ 1).$$

Similarly, we have

$$\begin{aligned} e_{path}(S_1 \to S_3) = & w_1 e_{path}(S_1 \to S_3 | Process\ 1) \\ & + w_2 e_{path}(S_1 \to S_3 | Process\ 3). \end{aligned}$$

In the next section, the above method is demonstrated. In general, as an extension of Definition 4.1, the following definition is made:

Definition 4.4 Let path $S_{i_1} \to S_{i_2} \to \cdots \to S_{i_J}$ be a partial path in (4.44). Then, the pathway effect of S_{i_1} on S_{i_J} is defined by

$$e_{path}\left(S_{i_1} \to S_{i_2} \to \cdots \to S_{i_J}\right) = \sum_{k=1}^{K} w_k e_{path}\left(S_{i_1} \to S_{i_2} \to \cdots \to S_{i_J} | Process\ k\right).$$

By using learning structure (4.35), the above definition is demonstrated. The path diagram of latent variables $S_i, i = 1, 2, 3, 4$ is illustrated in Fig. 4.4. For example, the pathway effect of $S_1 \to S_2 \to S_3$ is

$$e_{path}(S_1 \to S_2 \to S_3) = \sum_{k=1}^{3} w_k e_{path}(S_1 \to S_2 \to S_3 | Process\ k)$$

Fig. 4.4 Path diagram for learning structure (4.35)

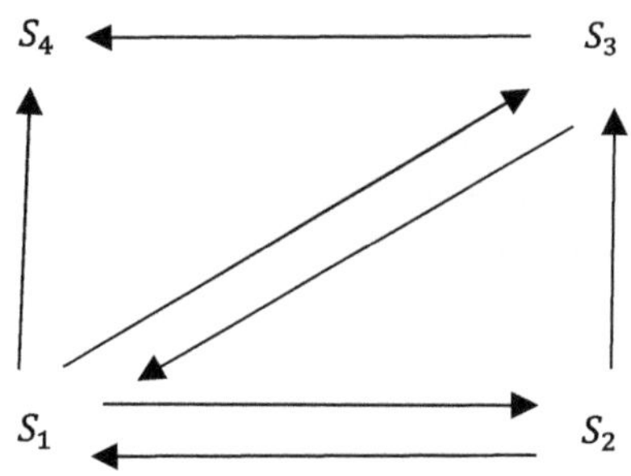

$$= w_1 e_{path}(S_1 \to S_2 \to S_3 | Process\ 1),$$

because Processes 2 and 3 do not have the path, i.e.,
$e_{path}(S_1 \to S_2 \to S_3 | Process\ 1) = e_{path}(S_1 \to S_2 \to S_3 | Process\ 3) = 0$.
Similarly, we have

$$\begin{aligned} e_{path}(S_3 \to S_1 \to S_4) &= \sum_{k=1}^{3} w_k e_{path}(S_3 \to S_1 \to S_4 | Process\ k) \\ &= w_3 e_{path}(S_1 \to S_2 \to S_3 | Process\ 3). \end{aligned}$$

The above method is demonstrated in a numerical example in the next section.

4.9 Numerical Illustration (Confirmatory Analysis)

In this section, a confirmatory analysis for explaining a learning structure is demonstrated by using the above discussion. Table 4.4 shows the first data set in Proctor [11]. For performing the analysis, let us assume there exist at most the following three learning processes in a population:

$$\begin{cases} Process\ 1(w_1) : S_5 \to S_4 \to S_3 \to S_2 \to S_1, \\ Process\ 2(w_2) : S_4 \to S_5 \to S_3 \to S_2 \to S_1, \\ Process\ 3(w_3) : S_4 \to S_3 \to S_5 \to S_2 \to S_1. \end{cases} \tag{4.46}$$

From the above learning processes, we have seven sub-structures. Let Structure (i) be the learning structures made by Process $(i), i = 1, 2, 3$; Structure (i, j) be the structures composed of Processes i and $j, i < j$; and let Structure $(1, 2, 3)$ be the structure formed by three processes in (4.46). Then, the skill acquisition patterns in the learning structures are illustrated in Table 4.5. From the table, Structures (1,2,3) and (1,3) have the same skill acquisition patterns, so in this sense, we cannot identify the structures from latent class model (4.3). The path diagram based on skill acquisition patterns $(s_1, s_2, s_3, s_4, s_5)$ is produced by learning Structure (1,2,3) (Fig. 4.5), so learning Structure (1,2,3) is an unclear one. On the other hand, learning Structure (1,3) produces a clear structure as shown in Fig. 4.6. In order to demonstrate the discussion in the previous section, latent class model (4.3) with skill acquisition patterns of learning Structure (1,2,3) is used. The results of the analysis are given in Table 4.6. First, based on the path diagram shown in Fig. 4.5, the following separating equations are obtained:

$$w_1 = w_1(v(0,0,0,0,0) + v(0,0,1,1,1) + v(0,1,1,1,1) + v(1,1,1,1,1)) + \frac{w_1}{w_1 + w_2} v(0,0,0,1,1) + v(0,0,0,0,1), \tag{4.47}$$

$$w_2 = w_2(v(0,0,0,0,0) + v(0,0,1,1,1) + v(0,1,1,1,1) + v(1,1,1,1,1)) + \frac{w_2}{w_2 + w_3} v(0,0,0,1,0) + \frac{w_2}{w_1 + w_2} v(0,0,0,1,1), \tag{4.48}$$

$$w_3 = w_3(v(0,0,0,0,0) + v(0,0,1,1,1) + v(0,1,1,1,1) + v(1,1,1,1,1)) + \frac{w_3}{w_2 + w_3} v(0,0,0,1,0) + v(0,0,1,1,0). \tag{4.49}$$

From the above equations, we have

Table 4.4 Proctor's first data

Response pattern	Frequency	Response pattern	Frequency
0 0 0 0 0	14	1 0 0 0 0	2
0 0 0 0 1	4	1 0 0 0 1	1
0 0 0 1 0	7	1 0 0 1 0	0
0 0 0 1 1	8	1 0 0 1 1	3
0 0 1 0 0	2	1 0 1 0 0	2
0 0 1 0 1	3	1 0 1 0 1	7
0 0 1 1 0	6	1 0 1 1 0	1
0 0 1 1 1	10	1 0 1 1 1	14
0 1 0 0 0	4	1 1 0 0 0	0
0 1 0 0 1	4	1 1 0 0 1	3
0 1 0 1 0	5	1 1 0 1 0	0
0 1 0 1 1	7	1 1 0 1 1	9
0 1 1 0 0	1	1 1 1 0 0	1
0 1 1 0 1	3	1 1 1 0 1	10
0 1 1 1 0	1	1 1 1 1 0	2
0 1 1 1 1	17	1 1 1 1 1	62

Data Source Proctor [11]

Table 4.5 Skill acquisition patterns for learning structures

Structure	Skill acquisition patterns
Structure (1)	(0,0,0,0,0), (0,0,0,0,1), (0,0,0,1,1), (0,0,1,1,1), (0,1,1,1,1), (1,1,1,1,1)
Structure (2)	(0,0,0,0,0), (0,0,0,1,0), (0,0,0,1,1), (0,0,1,1,1), (0,1,1,1,1), (1,1,1,1,1)
Structure (3)	(0,0,0,0,0), (0,0,0,1,0), (0,0,1,1,0), (0,0,1,1,1), (0,1,1,1,1), (1,1,1,1,1)
Structure (1,2)	(0,0,0,0,0), (0,0,0,1,0), (0,0,0,0,1), (0,0,0,1,1), (0,0,1,1,1), (0,1,1,1,1), (1,1,1,1,1)
Structure (2,3)	(0,0,0,0,0), (0,0,0,1,0), (0,0,0,1,1), (0,0,1,1,0), (0,0,1,1,1), (0,1,1,1,1), (1,1,1,1,1)
Structure (1,3)	(0,0,0,0,0), (0,0,0,1,0), (0,0,0,1,1), (0,0,1,1,0), (0,0,0,1,1), (0,0,1,1,1), (0,1,1,1,1), (1,1,1,1,1)
Structure (1,2,3)	(0,0,0,0,0), (0,0,0,1,0), (0,0,0,1,1), (0,0,1,1,0), (0,0,0,1,1), (0,0,1,1,1), (0,1,1,1,1), (1,1,1,1,1)

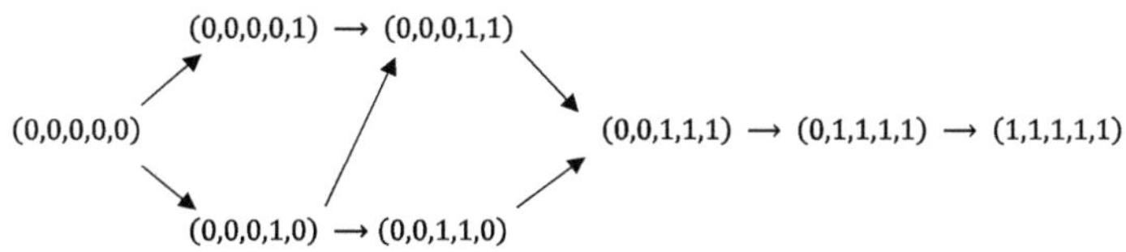

Fig. 4.5 Path diagram of skill acquisition patterns in learning Structure (1,2,3)

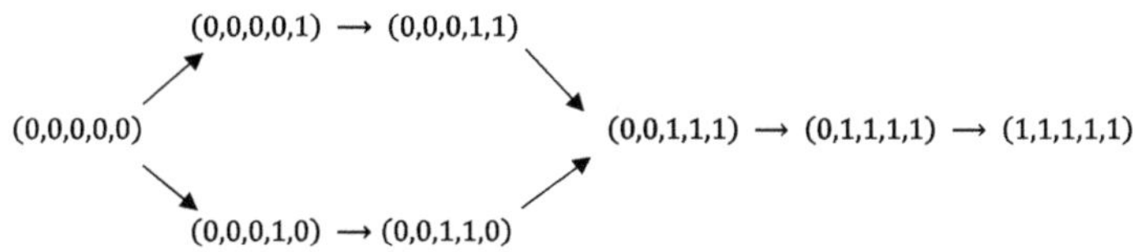

Fig. 4.6 Path diagram of skill acquisition patterns in learning Structure (1,3)

Table 4.6 The estimated positive response probabilities in learning Structure (1.2.3)

Skill acquisition pattern	Class proportion	Item response probability				
		Item 1	Item 2	Item 3	Item 4	Item 5
(0,0,0,0,0)	0.144	0.097	0.294	0.145	0.198	0.140
(0,0,0,0,1)	0.016	0.097	0.294	0.145	0.198	0.969
(0,0,0,1,0)	0.049	0.097	0.294	0.145	0.812	0.140
(0,0,0,1,1)	0.065	0.097	0.294	0.145	0.812	0.969
(0,0,1,1,0)	0.041	0.097	0.294	0.864	0.812	0.140
(0,0,1,1,1)	0.046	0.097	0.297	0.864	0.812	0.969
(0,1,1,1,1)	0.092	0.097	0.781	0.864	0.812	0.969
(1,1,1,1,1)	0.548	0.923	0.781	0.864	0.812	0.969

$G^2 = 16.499$, $(df = 15,\ P = 0.350)$

$$w_1 = \frac{v(0,0,0,0,1)}{v(0,0,0,1,0)+v(0,0,0,0,1)},$$

$$w_2 = 1 - w_1 - w_3,$$

$$w_3 = \frac{v(0,0,1,1,0)}{v(0,0,1,1,0)+v(0,0,0,1,1)}.$$

By using the estimates in Table 4.5, the estimates of the mixed proportions are calculated as follows:

$$\widehat{w}_1 = 0.246,\ \widehat{w}_2 = 0.368,\ \widehat{w}_3 = 0.386.$$

The above solution is a proper solution. The discussion in Sect. 4.5 is applied to this example. Let $v(s_1, s_2, s_3, s_4, s_5|Process\ k)$ be the proportions of individuals with skills $(s_1, s_2, s_3, s_4, s_5)$ in Process $k, k = 1, 2, 3$. For example, considering (4.47) for Process 1, we have

$$\begin{aligned} 1 = {} & v(0,0,0,0,0) + v(0,0,1,1,1) + v(0,1,1,1,1) \\ & + v(1,1,1,1,1) + \frac{1}{w_1 + w_2} v(0,0,0,1,1) + \frac{1}{w_1} v(0,0,0,0,1), \end{aligned}$$

so it follows that

$$v(0,0,0,0,0|Process\ 1) = v(0,0,0,0,0),$$

$$v(0,0,0,0,1|Process\ 1) = \frac{1}{w_1} v(0,0,0,0,1)$$

$$v(0,0,0,1,1|Process\ 1) = \frac{1}{w_1 + w_2} v(0,0,0,1,1),$$

$$v(0,0,1,1,1|Process\ 1) = v(0,0,1,1,1),$$

$$v(0,1,1,1,1|Process\ 1) = v(0,1,1,1,1),$$

$$v(1,1,1,1,1|Process\ 1) = v(1,1,1,1,1).$$

In Process 1 in (4.46), the sequence is a Markov chain and let $q^1_{11,i}$ be the related transition probabilities, for example, $q^1_{11,5}$ is related to path $S_5 \rightarrow S_4$ and $q^1_{11,4}$ to path $S_4 \rightarrow S_3$, and so on. Then, from Theorem 4.2, we have

$$q_{11,5}^{1}=\frac{\frac{1}{w_1+w_2}v(0,0,0,1,1)+v(0,0,1,1,1)+v(0,1,1,1,1)+v(1,1,1,1,1)}{\frac{1}{w_1}v(0,0,0,0,1)+\frac{1}{w_1+w_2}v(0,0,0,1,1)+v(0,0,1,1,1)+v(0,1,1,1,1)+v(1,1,1,1,1)},$$

$$q_{11,4}^{1}=\frac{v(0,0,1,1,1)+v(0,1,1,1,1)+v(1,1,1,1,1)}{\frac{1}{w_1+w_2}v(0,0,0,1,1)+v(0,0,1,1,1)+v(0,1,1,1,1)+v(1,1,1,1,1)},$$

$$q_{11,3}^{1}=\frac{v(0,1,1,1,1)+v(1,1,1,1,1)}{v(0,0,1,1,1)+v(0,1,1,1,1)+v(1,1,1,1,1)},$$

$$q_{11,2}^{1}=\frac{v(1,1,1,1,1)}{v(0,1,1,1,1)+v(1,1,1,1,1)}.$$

From the above results and Table 4.6, first, we have the following path coefficients:

$$Process\ 1\left(\widehat{w}_1=0.246\right): S_5 \overset{0.924}{\to} S_4 \overset{0.866}{\to} S_3 \overset{0.933}{\to} S_2 \overset{0.856}{\to} S_1.$$

Similarly, we get the estimates of Processes 2 and 3 as

$$Process\ 2\left(\widehat{w}_2=0.368\right): S_4 \overset{0.924}{\to} S_5 \overset{0.866}{\to} S_3 \overset{0.933}{\to} S_2 \overset{0.856}{\to} S_1,$$

$$Process\ 3\left(\widehat{w}_3=0.386\right): S_4 \overset{0.924}{\to} S_3 \overset{0.866}{\to} S_5 \overset{0.933}{\to} S_2 \overset{0.856}{\to} S_1.$$

According to Definition 4.3, second, the path coefficients of the above learning structure are calculated, for example, we have

$$e_{path}(S_5\to S_4)=w_1e_{path}(S_5\to S_4|Process\ 1)=0.246\times 0.924=0.227,$$

$$\begin{aligned}e_{path}(S_4\to S_3)&=w_1e_{path}(S_4\to S_3|Process\ 1)+w_3e_{path}(S_4\to S_3|Process\ 3)\\&=0.246\times 0.866+0.386\times 0.923=0.571.\end{aligned}$$

All the path coefficients calculated as above are illustrated in Fig. 4.7a. Third, some pathway effects in the learning structure are demonstrated. For example,

$$\begin{aligned}e_{path}(S_3\to S_2\to S_1)=&\ w_1e_{path}(S_3\to S_2\to S_1|Process\ 1)\\&+w_2e_{path}(S_3\to S_2\to S_1|Process\ 2)\\=&\ 0.246\times 0.933\times 0.856+0.358\times 0.933\times 0.856=0.340,\end{aligned}$$

$$\begin{aligned}e_{path}(S_3\to S_5\to S_2)&=w_3e_{path}(S_3\to S_5\to S_2|Process\ 3)\\&=0.386\times 0.866\times 0.933=0.312.\end{aligned}$$

$$e_{path}(S_5\to S_3\to S_2\to S_1)=w_2e_{path}(S_5\to S_2\to S_2\to S_1|Process\ 2)$$

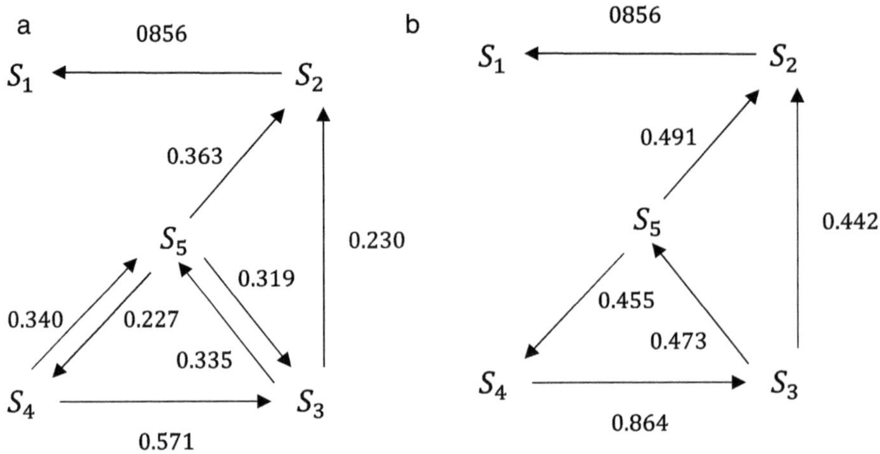

Fig. 4.7 **a** Path coefficients of learning structure (4.46). **b** Path coefficients of learning Structure (1,3)

$$= 0.368 \times 0.866 \times 0.933 \times 0.856 = 0.255,$$

and so on.

If the solution is improper, that is, there are negative estimates in the solution, Process 2 is deleted from the learning structure (4.46). Then, the learning structure becomes as follows:

$$\begin{cases} Process\ 1(w_1) : S_5 \to S_4 \to S_3 \to S_2 \to S_1 \\ Process\ 3(w_3) : S_4 \to S_3 \to S_5 \to S_2 \to S_1 \end{cases}, \tag{4.50}$$

and based on the path diagram shown in Fig. 4.6, we have

$$\begin{aligned} w_1 = {} & w_1(v(0,0,0,0,0) + v(0,0,1,1,1) + v(0,1,1,1,1) + v(1,1,1,1,1)) \\ & + v(0,0,0,1,1) + v(0,0,0,0,1), \end{aligned}$$

$$\begin{aligned} w_3 = {} & w_3(v(0,0,0,0,0) + v(0,0,1,1,1) + v(0,1,1,1,1) + v(1,1,1,1,1)) \\ & + v(0,0,0,1,0) + v(0,0,1,1,0). \end{aligned}$$

From the above equations, it follows that

$$w_1 = \frac{v(0,0,0,1,1) + v(0,0,0,0,1)}{v(0,0,0,1,1) + v(0,0,0,0,1) + v(0,0,0,1,0) + v(0,0,1,1,0)}, \ w_3 = 1 - w_1.$$

By using the estimates in Table 4.6, we obtain

$$\widehat{w}_1 = \frac{0.065 + 0.016}{0.065 + 0.016 + 0.049 + 0.041} = 0.474,\ \widehat{w}_3 = 1 - \widehat{w}_1 = 0.526.$$

In this learning structure, we have the following path coefficients:

$$\begin{cases} Process\ 1(\widehat{w}_1 = 0.474) : S_5 \overset{0.961}{\to} S_4 \overset{0.833}{\to} S_3 \overset{0.933}{\to} S_2 \overset{0.856}{\to} S_1. \\ Process\ 3(\widehat{w}_3 = 0.526) : S_4 \overset{0.891}{\to} S_3 \overset{0.898}{\to} S_5 \overset{0.933}{\to} S_2 \overset{0.856}{\to} S_{1.} \end{cases}$$

By using the above results, the path diagram of the skill acquisitions of $S_i, i = 1, 2, 3, 4, 5$ is illustrated in Fig. 4.7b.

4.10 A Method for Ordering Skill Acquisition Patterns

In this chapter, skill acquisition patterns, which are expressed as latent classes, are explained with a latent class model that is an extended version of the latent distance model discussed in Chap. 3. In the latent distance model, linear learning structures are discussed, so the skill acquisition patterns are naturally ordered; however, in the analysis of general learning structures, as treated in this chapter, such a natural ordering the skill acquisition patterns cannot be made, for example, learning structures in Figs. 4.5 and 4.6, skill acquisition patterns, $(0, 0, 0, 0, 1)$, $(0, 0, 0, 1, 1)$, $(0, 0, 0, 1, 0)$, $(0, 0, 1, 1, 0)$, cannot be order in a natural sense; however, it may be required to assess the levels with a manner. In this example, since it is clear pattern $(0, 0, 0, 0, 0)$ is the lowest and $(1, 1, 1, 1, 1)$ the highest, it is sensible to measure distances from $(0, 0, 0, 0, 0)$ or $(1, 1, 1, 1, 1)$ to skill acquisition patterns $(s_1, s_2, s_3, s_4, s_5)$ with a method. In order to do it, an entropy-based method for measuring distances between latent classes proposed in Chap. 2 (2.30) is used. In latent class model (4.3) with (4.4), let $\mathbf{0} = (0, 0, \ldots, 0)$ and $\boldsymbol{s} = (s_1, s_2, \ldots, s_I)$, and let $P(\boldsymbol{X} = \boldsymbol{x}|\mathbf{0})$ and $P(\boldsymbol{X} = \boldsymbol{x}|\boldsymbol{s})$ be the conditional distributions with the skill acquisition patterns, respectively. Then, from (2.30) the entropy-based distance between the skill acquisition patterns, i.e., latent classes, is defined by

$$D^*(P(\boldsymbol{X} = \boldsymbol{x}|\boldsymbol{s})||P(\boldsymbol{X} = \boldsymbol{x}|\mathbf{0})).$$

From the above distance, we have

$$\begin{aligned} D^*(P(\boldsymbol{X} = \boldsymbol{x}|\boldsymbol{s})\|P(\boldsymbol{X} = \boldsymbol{x}|\mathbf{0})) = \sum_{i=1}^{I} \{ & P(X_i = 1|0) \log \frac{P(X_i = 1|0)}{P(X_i = 1|s_i)} \\ & + (1 - P(X_i = 1|0)) \log \frac{1 - P(X_i = 1|0)}{1 - P(X_i = 1|s_i)} \\ & + P(X_i = 1|s_i) \log \frac{P(X_i = 1|s_i)}{P(X_i = 1|0)} \end{aligned}$$

$$+ (1 - P(X_i = 1|s_i)) \log \frac{1 - P(X_i = 1|s_i)}{1 - P(X_i = 1|0)}\}.$$

$$= \sum_{i=1}^{I} (P(X_i = 1|0) - P(X_i = 1|s_i)) \left(\log \frac{P(X_i = 1|0)}{1 - P(X_i = 1|0)} - \log \frac{P(X_i = 1|s_i)}{1 - P(X_i = 1|s_i)} \right), \tag{4.51}$$

where

$$P(X_i = 1|0) = P(X_i = 1|S_i = 0), P(X_i = 1|s_i) = P(X_i = 1|S_i = s_i).$$

Let

$$D^*(P(X_i = x_i|s_i)||P(X_i = x_i|0)) = (P(X_i = 1|0) - P(X_i = 1|s_i)) \left(\log \frac{P(X_i = 1|0)}{1 - P(X_i = 1|0)} - \log \frac{P(X_i = 1|s_i)}{1 - P(X_i = 1|s_i)} \right)$$

Then, the above quantity is an entropy-based distance between distributions $P(X_i = x_i|s_i)$ and $P(X_i = x_i|0)$. By using the notation, formula (4.51) becomes

$$D^*(P(\boldsymbol{X} = \boldsymbol{x}|\boldsymbol{s})||P(\boldsymbol{X} = \boldsymbol{x}|\boldsymbol{0})) = \sum_{i=1}^{I} D^*(P(X_i = x_i|s_i)||P(X_i = x_i|0))$$
$$= \sum_{i=1}^{I} s_i D^*(P(X_i = x_i|1)||P(X_i = x_i|0)). \tag{4.52}$$

In Fig. 4.6, let $\boldsymbol{0} = (0, 0, \ldots, 0)$ and $\boldsymbol{s} = (0, 0, 1, 1, 0)$, then, from (4.51) we have

$$D^*(P(\boldsymbol{X} = \boldsymbol{x}|\boldsymbol{s})||P(\boldsymbol{X} = \boldsymbol{x}|0)) = \sum_{i=3}^{4} D^*(P(X_i = x_i|1)||P(X_i = x_i|0)).$$

Applying the above method to Table 4.6, the skill acquisition patterns are ordered. Table 4.7 shows the entropy-based distances $D^*(P(X_i = x_i|1)||P(X_i = x_i|0))$, and by using the distances, we have the distances $D^*(P(\boldsymbol{X} = \boldsymbol{x}|\boldsymbol{s})||P(\boldsymbol{X} = \boldsymbol{x}|\boldsymbol{0}))$ which are in an increasing order (Table 4.8). For example, with respect to skill acquisition

Table 4.7 Entropy-based distances $D^*(P(X_i = x_i|1)||P(X_i = x_i|0))$ with respect to manifest variables X_i for Table 4.5

Manifest variable	X_1	X_2	X_3	X_4	X_5
$D^*(P(X_i = x_i\|1)\|\|P(X_i = x_i\|0))$	3.894	1.046	2.605	1.757	4.359

Table 4.8 Entropy-based distances $D^*(P(\boldsymbol{X} = \boldsymbol{x}|\boldsymbol{s})||P(\boldsymbol{X} = \boldsymbol{x}|0))$ with respect to skill acquisition patterns for Table 4.5

Skill acquisition pattern	$D^*(P(\boldsymbol{X} = \boldsymbol{x}|\boldsymbol{s})||P(\boldsymbol{X} = \boldsymbol{x}|0))$
(0, 0, 0, 0, 0)	0
(0, 0, 0, 1, 0)	1.757
(0, 0, 0, 0.1)	4.359
(0, 0, 1, 1, 0)	4.362
(0, 0, 0, 1, 1)	6.116
(0, 0, 1, 1, 1)	8.721
(0, 1, 1, 1, 1)	9.767
(1, 1, 1, 1, 1)	13.661

patterns (0, 0, 1, 1, 0) and (0, 0, 0, 1, 1), the latter can be regarded as a higher level than the former.

Remark 4.3 In the above method for grading the latent classes (skill acquisition patterns), we can use the distances from $\mathbf{1} = (1, 1, \ldots, 1)$ as well. Then, it follows that

$$\begin{aligned} &D^*(P(\boldsymbol{X} = \boldsymbol{x}|\boldsymbol{s})||P(\boldsymbol{X} = \boldsymbol{x}|\mathbf{1})) \\ &\quad = D^*(P(\boldsymbol{X} = \boldsymbol{x}|\mathbf{1})||P(\boldsymbol{X} = \boldsymbol{x}|\mathbf{0})) - D^*(P(\boldsymbol{X} = \boldsymbol{x}|\boldsymbol{s})||P(\boldsymbol{X} = \boldsymbol{x}|\mathbf{0})). \end{aligned}$$

For example, in Table 4.8, for $\boldsymbol{s} = (0, 0, 1, 1, 1)$, we have

$$D^*(P(\boldsymbol{X} = \boldsymbol{x}|\boldsymbol{s})||P(\boldsymbol{X} = \boldsymbol{x}|1)) = 13.661 - 8.721 = 4.940.$$

Hence, the results from the present ordering (grading) method based on $\mathbf{1} = (1, 1, \ldots, 1)$ are intrinsically the same as that based on $\mathbf{0} = (0, 0, \ldots, 0)$. ■

In latent class model (4.3), let $\boldsymbol{s}_1 = (s_{11}, s_{12}, \ldots, s_{1I})$ and $\boldsymbol{s}_2 = (s_{21}, s_{22}, \ldots, s_{2I})$ be skill acquisition patterns. From (4.50) and (4.51), the difference between the two skill acquisition patterns, i.e., latent classes, is calculated by

$$D^*(P(\boldsymbol{X} = \boldsymbol{x}|\boldsymbol{s}_1)||P(\boldsymbol{X} = \boldsymbol{x}|\boldsymbol{s}_2)) = \sum_{i=1}^{I} |s_{1i} - s_{2i}| D^*(P(X_i = x_i|1)||P(X_i = x_i|0)). \tag{4.53}$$

For example, in the example shown in Tables 4.7 and 4.8, for $\boldsymbol{s}_1 = (0, 0, 1, 1, 0)$ and $\boldsymbol{s}_2 = (0, 1, 1, 1, 1)$, the difference between the latent classes is calculated as follows:

$$\begin{aligned} D^*(P(\boldsymbol{X} = \boldsymbol{x}|\boldsymbol{s}_1)||P(\boldsymbol{X} = \boldsymbol{x}|\boldsymbol{s}_2)) &= D^*(P(X_2 = x_2|1)||P(X_2 = x_2|0)) \\ &\quad + D^*(P(X_5 = x_5|1)||P(X_5 = x_5|0)) \end{aligned}$$

$$= 1.046 + 4.359 = 5.405.$$

The difference calculated above can be interpreted as the distance between the latent classes, measured in entropy. Figure 4.8 shows an undirected graph and the values are the entropy-based difference between the latent classes. The distance between the latent classes can be calculated by summing the values in the shortest way between the latent classes, for example, in Fig. 4.8, there are two shortest ways between (0, 0, 0, 1, 0) and (0, 0, 1, 1, 1):

(i) (0,0,0,1,0) ----- (0,0,0,1,1) ----- (0,0,1,1,1),
(ii) (0,0,0,1,0) ----- (0,0,1,1,0) ----- (0,0,1,1,1).

By the first way, the distance is calculated as $4.359 + 2.605 = 6.964$, and the same result is also obtained from the second way. It may be significant to make a tree graph of latent classes by using cluster analysis with entropy (Chap. 2, Sect. 5), in order to show the relationship of the latent classes. From Fig. 4.8, we have a tree graph of the latent classes (Fig. 4.9).

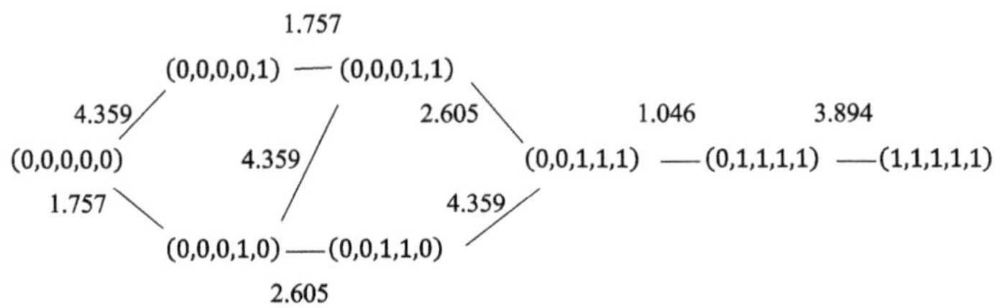

Fig. 4.8 Undirected graph for explaining the differences between the latent classes

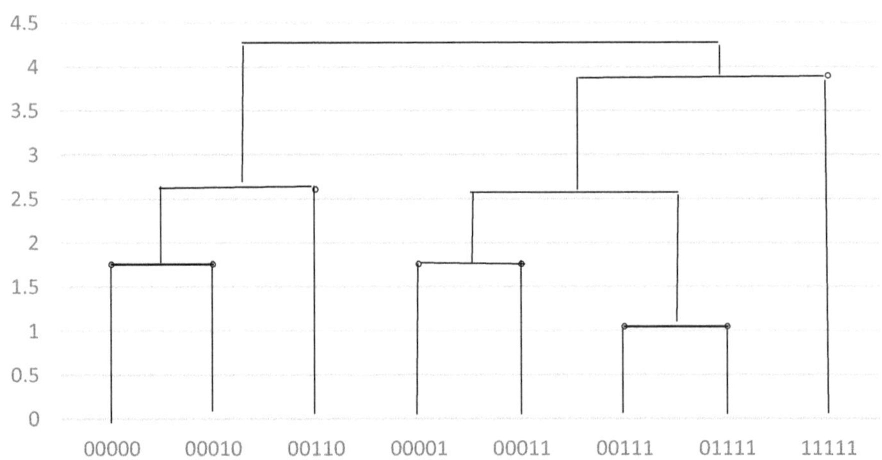

Fig. 4.9 A tree graph of latent classes in Table 4.6 based on entropy

Table 4.9 Entropy-based distances $D^*(P(X_i = x_i|1)||P(X_i = x_i|0))$ with respect to manifest variables X_i for Table 3.8 in Chap. 3

Manifest variable	X_{11}	X_{12}	$X1_3$	X_{21}	X_{22}	X_{23}
$D^*(P(X_i = x_i\|1)\|\|P(X_i = x_i\|0))$	3.193	4.368	3.637	4.413	2.855	5.543

Table 4.10 Entropy-based distances $D^*(P(\boldsymbol{X} = \boldsymbol{x}|\boldsymbol{s})||P(\boldsymbol{X} = \boldsymbol{x}|0))$ with respect to skill acquisition patterns for Table 3.8 in Chap. 3

Latent class	$D^*(P(\boldsymbol{X} = \boldsymbol{x}\|\boldsymbol{s})\|\|P(\boldsymbol{X} = \boldsymbol{x}\|0))$	Latent class	$D^*(P(\boldsymbol{X} = \boldsymbol{x}\|\boldsymbol{s})\|\|P(\boldsymbol{X} = \boldsymbol{x}\|0))$
(0, 0)	0	(0, 2)	8.398
(1, 0)	3.367	(1, 2)	12.036
(2, 0)	8.005	(2, 2)	16.404
(3, 0)	11.198	(3, 2)	19.596
(0, 1)	5.543	(0, 3)	12.812
(1, 1)ara>	9.180	(1, 3)	16.449
(2, 1)	13.549	(2, 3)	20.817
(3, 1)	16.741	(3, 3)	24,010

The present method for grading latent classes is applied to an example treated in Sect. 3.4 (Chap. 3). The estimated latent classes are expressed as score vectors $(i, j), i = 0, 1, 2, 3; j = 0, 1, 2, 3$, which imply pairs of levels for the general intelligence θ_1 and the verbal ability of children θ_2, respectively. In this example, although the levels of child ability can be graded according to the sum of the scores, $t = i + j$, it may be meaningful to use the present method for the grading. From the estimated model shown in Table 3.8, we have the entropy-based distances with respect to manifest variables (Table 4.9). Distances $D^*(P(\boldsymbol{X} = \boldsymbol{x}|\boldsymbol{s})||P(\boldsymbol{X} = \boldsymbol{x}|0))$ are calculated in Table 4.10, where $0 = (0, 0)$. By using the distances, grading of the latent classes can be made. For example, for score $t = 3$, the order of latent classes (3, 0), (2, 1) (1, 2), and (0.3) is as follows.

$$(3, 0) < (1.2) < (0, 3) < (2, 1).$$

4.11 Discussion

The present chapter has applied latent class analysis to explain learning structures. Skill acquisitions are scaled with the related test items (manifest variables), and the states of skill acquisition are expressed by latent binary variables, and thus, manifest responses measure the states with response errors, i.e., omission (forgetting) and intrusion (guessing) ones. The structures expressed in this context are called the learning structures in this book. When the skills under consideration are ordered with respect to prerequisite relationships, for example, for skills in calculation, (1) addition, (2) multiplication, and (3) division, the learning structure is called a linear learning structure. The model in this chapter is an extension of the latent distance model. From the learning structure, the traces of skill learning process in a population can be discussed through the path diagrams of skill acquisition patterns, and based on the traces, dynamic interpretations of the learning structures can be made. In general, learning structures are not necessarily linear, that is, there exist some learning processes of skills in a population. Hence, it is valid to assume that the population is divided into several subpopulations that depend on learning processes of their own. The present chapter gives a method to explain learning processes of skills by using cross-sectional data. It is assumed that manifest variables depend only on the corresponding latent variables (states of skills); however, it is more realistic to introduce "transfer effects" in the latent class models [1, 2]. In the above example of skills for calculation, it is easily seen that the skill of addition is prerequisite to that of multiplication. In this case, the mastery of the skill of multiplication will facilitate the responses to test items for addition, i.e., a "facilitating" transfer effect of multiplication on addition, and thus, it is more appropriate to take the transfer effects into account to discuss learning structures. In the other way around, there may be cases where "inhibiting" transfer effects are considered in analysis of learning structures [9]. In this chapter, "transfer effects" have not been hypothesized on the latent class model. It is significant to go into further studies to handle the transfer effects as well as prerequisite relationships between skills in studies on learning. Approaches to pairwise assessment of prerequisite relationships between skills were made by several authors, for example, White and Clark [12], Macready [9], and Eshima et al. [5]. It is the first attempt that Macready [7] dealt with the transfer effects in a pairwise assessment of skill acquisition by using latent class models with equality constraints. In order to improve the model, Eshima et al. [5] proposed a latent class model structured with skill acquisition and transfer effect parameters for making pairwise assessments of kill acquisition; however, transfer effect parameters in the model are common to the related manifest variables. The study on the pairwise assessment of prerequisite relationships among skills is important to explain learning structures, and the studies based on latent structure models are left as significant themes in the future.

References

1. Bergan, J. R. (1980). The structural analysis of behavior: An alternative to the learning-hierarchy model. *Review of Educational Research, 50*, 625–646.
2. Coptton, J. W., Gallagher, J. P., & Marshall, S. P. (1977). The identification and decomposition of hierarchical tasks. *American Educational Research Journal, 14*, 189–212.
3. Dayton, M., & Macready, G. B. (1976). A probabilistic model for validation of behavioral hierarchies. *Psychometrika, 41*, 190–204.
4. Dayton, M., & Macready, G. B. (1980). A scaling model with response errors and intrinsically unscalable respondents. *Psychometrika*, 344–356.
5. Eshima, N. (1990). Latent class analysis for explaining a hierarchical learning structure. *Journal of the Japan Statistical Society, 20*, 1–12.

6. Eshima, N., Asano, C., & Tabata, M. (1996). A developmental path model and causal analysis of latent dichotomous variables. *British Journal of Mathematical and Statistical Psychology, 49*, 43–56.
7. Eshima, N., Asano, C., & Obana, E. (1990). A latent class model for assessing learning structures. *Behaviormetrika, 28*, 23–35.
8. Goodman, L. A. (1975). A new model for scaling response patterns: An application of quasi-independent concept. *Journal of the American Statistical Association, 70*, 755–768.
9. Macready, G. B. (1982). The use of latent class models for assessing prerequisite relations and transference among traits. *Psychometrika, 47*, 477–488.
10. Magidson, J., & Vermunt, J. K. (2001). Latent class factor and cluster models: Bi-plots, and related graphical displays. *Sociological Methodology, 31*, 223–264.
11. Proctor, C. H. (1970). A probabilistic formulation and statistical analysis of Guttman scaling. *Psychometrika, 35*, 73–78.
12. White, R. T., & Clark, R. M. (1973). A test of inclusion which allows for errors of measurement. *Psychometrika, 38*, 77–86.

Chapter 5
The Latent Markov Chain Model

5.1 Introduction

The Markov chain model is important for describing time-dependent changes of states in human behavior; however, when observing changes of responses to a question about a particular characteristic, individuals' responses to the question may not reflect the true states of the characteristic. As an extension of the Markov chain model, the latent Markov chain model was proposed in an unpublished Ph.D. dissertation by Wiggins L. M. in 1955 [5, 14]. The assumptions of the model are (i) at every observed time point, a population is divided into several latent states, which are called latent classes as well in the present chapter; (ii) an individual in the population takes one of the manifest states according to his or her latent state at the time point; and (iii) the individual changes the latent states according to a Markov chain. The assumptions are the same as those of the hidden Markov model in time series analysis [6, 7]. In behavioral sciences, for individuals in a population, responses to questions about particular characteristics may be observed several times to explain the changes of responses. In this case, the individuals' responses to the questions are viewed as the manifest responses that may not reflect their true states at the observed time points, that is, intrusion and omission errors have to be taken into consideration. The response categories to be observed are regarded as the manifest states and the true states of the characteristics, which are not observed directly, as the latent states. Concerning parameter estimation in the latent Markov chain model, algebraic methods were studied by Katz and Proctor [15] and [14]. The methods were given for cases where the number of manifest states equals that of latent states, so were not able to treat general cases. Moreover, the methods may derive improper estimates of transition probabilities, for example, negative estimates of the probabilities, and any method of assessing the goodness-of-fit of the model to data sets was not given. These shortages hinder the application of the model to practical research in behavioral sciences. The Markov chain model is a discrete-time model, and may be an asymptotic one for the continuous-time model. In most social or behavioral phenomena, an

N. Eshima, *An Introduction to Latent Class Analysis*, Behaviormetrics: Quantitative Approaches to Human Behavior 14,
https://doi.org/10.1007/978-981-19-0972-6_5

individual in a population changes his or her states in continuous time; however, it is difficult to observe the changes continuously, and so data are collected, for example, monthly or annually. Thus, it is significant to describe the changes with discrete-time models, in which changes are treated as if they took place at the observed time points. Continuous-time models are also important to explain changes in human behaviors, so discussions for the parameter estimations were made by Singer and Spilerman [2, 16–18], and so on. In the present chapter, the discrete-time model is called the latent Markov chain model, and the continuous-time one the latent Markov process model, for convenience.

In the present chapter, the latent Markov models are discussed and an ML estimation procedure is made via the EM algorithm. In Sect. 5.2, the latent Markov chain model is explained and the relationship between the usual latent class model and the Markov model is considered. Section 5.3 constructs an ML estimation procedure for the latent Markov chain model via the EM algorithm, and Sect. 5.4 gives a property of the procedure, preferable for the parameter estimation. In Sects. 5.5 and 5.6, numerical examples are given to demonstrate the ML estimation procedure. Section 5.7 considers an example of the latent Markov chain model missing manifest observations, and Sect. 5.8 discusses a general model for treating such cases. In Sect. 5.9, the latent Markov process model with finite manifest and latent states is considered. Finally, in Sect. 5.10, discussions and themes to be studied through further research are given.

5.2 The Latent Markov Chain Model

Let X_t be manifest variables that take values on sample space $\Omega_{manifest} = \{1, 2, \ldots, J\}$ at time points $t = 1, 2, \ldots$, and let S_t be the corresponding latent variables on sample space $\Omega_{latent} = \{1, 2, \ldots, A\}$. In what follows, states on $\Omega_{manifest}$ are called manifest states and those on Ω_{laten} latent ones. At time point t, it is assumed an individual in a population takes a manifest state on $\Omega_{manifest}$ according to his latent state on Ω_{latent} and he changes the latent states according to a (first-order) Markov chain $S_t, t = 1, 2, \ldots$. First, the Markov chain is assumed to be time-homogeneous, that is, the transition probabilities are independent of time points. Let $m_{ab}, a, b = 1, 2, \ldots, A$ be the transition probabilities; let $v_a, a = 1, 2, \ldots, A$ be the probabilities of $S_1 = a$, that is, the initial state distribution; and let p_{aj} be the probabilities of $X_t = j$, given $S_t = a$, that is, $p_{aj} = P(X_t = j|S_t = a)$, and let $p(x_1, x_2, \ldots, x_T)$ be the probabilities with which an individual takes manifest state transition $x_1 \to x_2 \to \cdots \to x_T$. Then, the following accounting equations can be obtained:

$$p(x_1, x_2, \ldots, x_T) = \sum_{s} v_{s_1} p_{s_1 x_1} \prod_{t=1}^{T-1} m_{s_t s_{t+1}} p_{s_{t+1} x_{t+1}}, \quad (5.1)$$

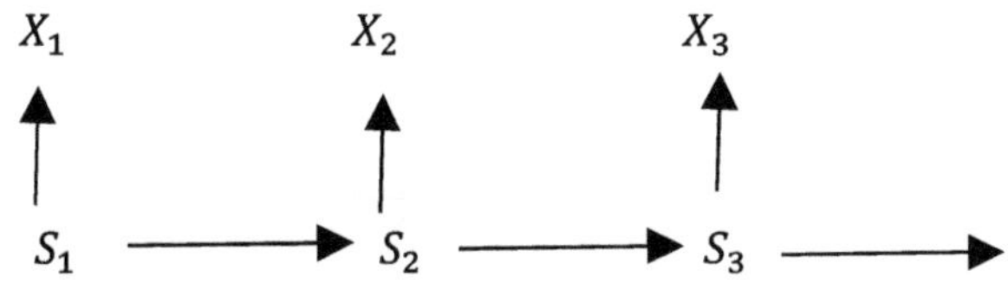

Fig. 5.1 Path diagram of the latent Markov chain model

where the summation in the above equations implies that over all latent states $s = (s_1, s_2, \ldots, s_T)$. The parameters are restricted as

$$\sum_{a=1}^{A} v_a = 1, \sum_{b=1}^{A} m_{ab} = 1, \sum_{x=1}^{J} p_{ax} = 1. \tag{5.2}$$

The above equations specify the time-homogeneous latent Markov chain model that is an extension of the Markov chain model, which is expressed by setting $A = J$ and $p_{aa} = 1, a = 1, 2, \ldots, A$. The Markov chain model is expressed as

$$p(x_1, x_2, \ldots, x_T) = v_{x_1} \prod_{t=1}^{T-1} m_{x_t x_{t+1}}.$$

For the latent Markov chain model, the path diagram of manifest variables X_t and latent variables S_t is illustrated in Fig. 5.1.

Second, the non-homogeneous model, that is, the latent Markov chain model with non-stationary transition probabilities, is treated. Let $m_{(t)ab}, a, b = 1, 2, \ldots, A$ be transition probabilities at time point $t = 1, 2, \ldots, T - 1$. Then, the accounting equations are given by

$$p(x_1, x_2, \ldots, x_T) = \sum_{s} v_{s_1} p_{s_1 x_1} \prod_{t=1}^{T-1} m_{(t)s_t s_{t+1}} p_{s_{t+1} x_{t+1}}, \tag{5.3}$$

where

$$\sum_{a=1}^{A} v_a = 1, \sum_{b=1}^{A} m_{(t)ab} = 1, \sum_{x=1}^{J} p_{ax} = 1. \tag{5.4}$$

In the above model, it is assumed that the manifest response probabilities p_{aj} are independent of time points. If the probabilities depend on the observed time points, the probabilities are expressed as $p_{(t)ax}$, and then, the above accounting equations are modified as follows:

$$p(x_1, x_2, \ldots, x_T) = \sum_{s} v_{s_1} p_{(1)s_1 x_1} \prod_{t=1}^{T-1} m_{(t)s_t s_{t+1}} p_{(t+1)s_{t+1} x_{t+1}}, \tag{5.5}$$

where

$$\sum_{a=1}^{A} v_a = 1, \sum_{b=1}^{A} m_{(t)ab} = 1, \sum_{x=1}^{J} p_{(t)ax} = 1. \tag{5.6}$$

In the above model, set $m_{(t)ab} = 0, a \neq b$, then, (5.5) becomes

$$p(x_1, x_2, \ldots, x_T) = \sum_{a=1}^{A} v_a \prod_{t=1}^{T} p_{(t)ax_t}. \tag{5.7}$$

In the above expression, regarding variables $X_t, t = 1, 2, \ldots, T$ as T item responses formally, then, the above equations are those for the usual latent class model with A latent classes. In this sense, the latent Markov chain model is an extension of the usual latent class model. On the contrary, since (5.5) can be reformed as

$$p(x_1, x_2, \ldots, x_T) = \sum_{S} v_{s_1} \prod_{t=1}^{T-1} m_{(t)s_t s_{t+1}} \prod_{t=1}^{T} p_{(t)s_t x_t}, \tag{5.8}$$

regarding latent state transition patterns in $\Pi_{t=1}^{T} \Omega_{latent} = \overbrace{\Omega_{latent} \times \Omega_{latent} \cdots \times \Omega_{latent}}^{T}$, from (5.8) the class proportions with latent state transitions $s_1 \to s_2 \to \cdots \to s_T$ are given by $v_{s_1} \prod_{t=1}^{T-1} m_{(t)s_t s_{t+1}}$. Hence, the above discussion derives the following theorem.

Theorem 5.1 The latent class model (1.2) and the latent Markov chain model (5.5) are equivalent. ■

Remark 5.1 The latent Markov chain models treated above have responses to one question (manifest variable) at each observed time point. Extended versions of the models can be constructed by introducing a set of questions (a manifest variable vector) $\boldsymbol{X} = (X_1, X_2, \ldots, X_I)$. For the manifest variable vector, responses are observed as

$$\boldsymbol{x}_1 \to \boldsymbol{x}_2 \to \cdots \to \boldsymbol{x}_T,$$

where

$$\boldsymbol{x}_t = (x_{t1}, x_{t2}, \ldots, x_{tI}), t = 1, 2, \ldots, T.$$

Setting $T = 1$, the above model is the usual latent class model (1.2). Let $p_{is_t x_{ti}}, i = 1, 2, \ldots, I$ be the response probabilities for manifest variables $\boldsymbol{X}_t =$

$(X_1, X_2, \ldots, X_I)$, given latent state $S_t = s_t$, that is, $\prod_{i=1}^{I} p_{is_t x_{ti}}$ at time points $t = 1, 2, \ldots, T$. Then, model (5.5) is extended as

$$p(\boldsymbol{x}_1, \boldsymbol{x}_2, \ldots, \boldsymbol{x}_T) = \sum_{s} v_{s_1} \prod_{i=1}^{I} p_{is_1 x_{1i}} \prod_{t=1}^{T-1} \left(m_{(t)s_t s_{t+1}} \prod_{i=1}^{I} p_{is_{t+1} x_{t+1i}} \right), \tag{5.9}$$

where

$$\sum_{a=1}^{A} v_a = 1, \sum_{b=1}^{A} m_{(t)ab} = 1, \sum_{x=1}^{J} p_{iax} = 1,$$

and notation $\sum_{s}$ implies the summation over all latent state transitions $\mathbf{s} = (s_1, s_2, \ldots, s_T)$. An ML estimation procedure via the EM algorithm can be built with a method similar to the above ones. The above model is related to a multivariate extension of the Latent Markov chain model with covariates by Bartolucci and Farcomeni [3].

5.3 The ML Estimation of the Latent Markov Chain Model

First, the EM algorithm is considered for model (5.1) with constraints (5.2). Let $n(x_1, x_2, \ldots, x_T)$ be the numbers of individuals who take manifest state transitions (responses) $x \to x_2 \to \cdots \to x_T$; let $n(x_1, x_2, \ldots, x_T; s_1, s_2, \ldots, s_T)$ be those with the manifest state transitions and latent state transitions, $s_1 \to s_2 \to \cdots \to s_T$; and let N be the total of the observed individuals. Then,

$$\begin{cases} n(x_1, x_2, \ldots, x_T) = \sum_{s} n(x_1, x_2, \ldots, x_T; s_1, s_2, \ldots, s_T) \\ N = \sum_{x} n(x_1, x_2, \ldots, x_T) \end{cases},$$

where $\sum_{s}$ and $\sum_{x}$ imply summations over $\boldsymbol{s} = (s_1, s_2, \ldots, s_T) \in \prod_{t=1}^{T} \Omega_{latent}$ and $\boldsymbol{x} = (x_1, x_2, \ldots, x_T) \in \prod_{t=1}^{T} \Omega_{manifest}$, respectively. In this model, the complete and incomplete data are expressed by sets $D_{complete} = \{n(x_1, x_2, \ldots, x_T; s_1, s_2, \ldots, s_T)\}$ and $D_{incomplete} = \{n(x_1, x_2, \ldots, x_T)\}$, respectively. Let $\boldsymbol{\varphi} = \left((v_a), (m_{(t)ab}), (p_{ax})\right)$ be the parameter vector. Then, we have the following log likelihood function of $\boldsymbol{\varphi}$, given the complete data:

$$l(\boldsymbol{\varphi}|D_{incomplete}) = \sum_{\boldsymbol{x},\boldsymbol{s}} n(x_1, x_2, \ldots, x_T; s_1, s_2, \ldots, s_T) \left\{ \log v_{s_1} + \sum_{t=1}^{T-1} \log m_{s_t s_{t+1}} + \sum_{t=1}^{T} \log p_{s_t x_t} \right\}, \tag{5.10}$$

where $\sum_{x,s}$ implies the summation over manifest and latent states transition patterns $\boldsymbol{x} = (s_1, s_2, \ldots, s_T)$ and $\boldsymbol{s} = (s_1, s_2, \ldots, s_T)$. The model parameters $\boldsymbol{\varphi}$ are estimated by the EM algorithm. Let ${}^r\boldsymbol{\varphi} = ({}^r v_a), ({}^r m_{ab}), ({}^r p_{ax}))$ be the estimates at the r th iteration in the M-step; and let ${}^{r+1}D_{complete} = \left\{{}^{r+1}n(x_1, x_2, \ldots, x_T; s_1, s_2, \ldots, s_T)\right\}$ be the conditional expectations of the complete data $D_{complete}$ at the $r+1$ th iteration in the E-step. Then, the E- and M-steps are given as follows.

(i) E-step

In this step, the conditional expectation of (5.10) given parameters ${}^r\boldsymbol{\varphi}$ and $D_{incomplete}$ is calculated, that is,

$$Q(\boldsymbol{\varphi}|{}^r\boldsymbol{\varphi}) = E\left(l\left(\boldsymbol{\varphi}|D_{complete}\right)|{}^r\boldsymbol{\varphi}, D_{incomplete}\right). \tag{5.11}$$

Since the complete data are sufficient statistics, the step is reduced to calculating the conditional expectations of the complete data $n(x_1, x_2, \ldots, x_T; s_1, s_2, \ldots, s_T)$ and we have

$${}^{r+1}n(x_1, x_2, \ldots, x_T; s_1, s_2, \ldots, s_T) = n(x_1, x_2, \ldots, x_T)\frac{{}^r v_{s_1}\prod_{t=1}^{T-1}{}^r m_{s_t s_{t+1}}\prod_{t=1}^{T}{}^r p_{s_t x_t}}{\sum_{\boldsymbol{s}}{}^r v_{s_1}\prod_{t=1}^{T-1}{}^r m_{s_t s_{t+1}}\prod_{t=1}^{T}{}^r p_{s_t x_t}}. \tag{5.12}$$

(ii) M-step

Function (5.11) is maximized with respect to parameters v_{s_1}, $m_{s_t s_{t+1}}$, and $p_{s_t x_t}$ under constraints (5.2). By using Lagrange multipliers, $\kappa, \lambda_a, a = 1, 2, \ldots, A$ and $\mu_c, c = 1, 2, \ldots, A$, the Lagrange function is given by

$$L = Q(\boldsymbol{\varphi}|{}^r\boldsymbol{\varphi}) - \kappa\sum_{a=1}^{A} v_a - \sum_{a=1}^{A}\lambda_a\sum_{b=1}^{A} m_{ab} - \sum_{c=1}^{A}\mu_c\sum_{x=1}^{J} p_{ax}. \tag{5.13}$$

From the following equations and constraints (5.2)

$$\frac{\partial L}{\partial v_a} = 0, a = 1, 2, \ldots, A; \frac{\partial L}{\partial m_{ab}} = 0, a, b = 1, 2, \ldots, A;$$

$$\frac{\partial L}{\partial p_{ax}} = 0, a = 1, 2, \ldots, A, x = 1, 2, \ldots, J,$$

we have the following estimates:

$${}^{r+1}v_a = \frac{\sum_{x,s\backslash 1}{}^{r+1}n(x_1, x_2, \ldots, x_T; a, s_2, \ldots, s_T)}{N}, a = 1, 2, \ldots, A, \tag{5.14}$$

where $\sum_{x,s^{\setminus 1}}$ implies the summation over all $\boldsymbol{x} = (x_1, x_2, \ldots, x_T)$ and $\boldsymbol{s}^{\setminus 1} = (s_2, s_3, \ldots, s_T)$;

$$
{}^{r+1}m_{ab} = \frac{\sum_{t=1}^{T-1}\sum_{x,s^{\setminus t,t+1}} {}^{r+1}n(x_1, x_2, \ldots, x_T; s_1, s_2, \ldots, s_{t-1}, a, b, s_{t+2}, \ldots, s_T)}{\sum_{t=1}^{T-1}\sum_{x,s^{\setminus t}} {}^{r+1}n(x_1, x_2, \ldots, x_T; s_1, s_2, \ldots s_{t-1}, a, s_{t+1}, \ldots, s_T)},
$$
$$
a, b = 1, 2, \ldots, A, \tag{5.15}
$$

where $\sum_{x,s^{\setminus t}}$ implies the summation over all $\boldsymbol{x} = (x_1, x_2, \ldots, x_T)$ and $\boldsymbol{s}^{\setminus t} = (s_1, s_2, \ldots, s_{t-1}, s_{t+1}, \ldots, s_T)$;

$$
{}^{r+1}p_{ab} = \frac{\sum_{t=1}^{T-1}\sum_{,x^{\setminus t},s^{\setminus t}} {}^{r+1}n(x_1, x_2, \ldots, x_{t-1}, b, x_{t+1}, \ldots x_T; s_1, s_2, \ldots, s_{t-1}, a, s_{t+1}, \ldots, s_T)}{\sum_{t=1}^{T-1}\sum_{,x,,s^{\setminus t},} {}^{r+1}n(x_1, x_2, \ldots, x_{t-1}, x_t, x_{t+1}, \ldots x_T; s_1, s_2, \ldots, s_{t-1}, a, s_{t+1}, \ldots, s_T)}
$$
$$
a = 1, 2, \ldots, A;\ b = 1, 2, \ldots, J, \tag{5.16}
$$

where $\sum_{,x^{\setminus t},s^{\setminus t}}$ implies the summation over all $\boldsymbol{x} = (x_1, x_2, \ldots, x_{t-1}, x_{t+1}, \ldots, x_T)$ and $\boldsymbol{s}^{\setminus t} = (s_1, s_2, \ldots, s_{t-1}, s_{t+1}, \ldots, s_T)$.

Remark 5.2 In order to estimate the parameters of Markov models explained in Sect. 5.2, the conditions of model identification need to hold. For model identification of model (5.1) with constraints (5.2), the following condition is needed:

$$
The\,number\,of\ p(x_1, x_2, \ldots, x_T) > those\,of\ v_a, m_{ab}\,and\ p_{ax}. \tag{5.17}
$$

According to constraints (5.2) and

$$
\sum_{x} p(x_1, x_2, \ldots, x_T) = 1,
$$

constraint (5.17) becomes

$$
J^T - 1 > (A - 1) + A(A - 1) + A(J - 1) = A(A + J - 1) - 1
$$

$$
\Leftrightarrow J^T - A(A + J - 1) > 0 \tag{5.18}
$$

Similarly, for the other models, the model identification conditions can be derived. ■

Second, an ML estimation procedure for model (5.3) with constraints (5.4) is given. Let ${}^r\boldsymbol{\varphi} = \left(({}^r v_a), ({}^r m_{(t)ab}), ({}^r p_{ax})\right)$ be the estimates at the rth iteration in the M-step; and let ${}^{r+1}D_{complete} = \left\{{}^{r+1}n(x_1, x_2, \ldots, x_T | s_1, s_2, \ldots, s_T)\right\}$ be the conditional expectations of the complete data $D_{complete}$ at the $r + 1$ th iteration in the E-step. Then, the E- and M-steps are given as follows.

(i) E-step

$$
\begin{aligned}
&{}^{r+1}n(x_1, x_2, \ldots, x_T; s_1, s_2, \ldots, s_T) \\
&\quad = n(x_1, x_2, \ldots, x_T) \frac{{}^r v_{s_1} \prod_{t=1}^{T-1} {}^r m_{(t)s_t s_{t+1}} \prod_{t=1}^{T} {}^r p_{s_t x_t}}{\sum_{\boldsymbol{s}} {}^r v_{s_1} \prod_{t=1}^{T-1} {}^r m_{(t)s_t s_{t+1}} \prod_{t=1}^{T} {}^r p_{s_t x_t}}.
\end{aligned} \tag{5.19}
$$

(ii) M-step

Estimates ${}^{r+1}v_a$ and ${}^{r+1}p_{ax}$ are given by (5.14) and (5.16), respectively. We have ${}^{r+1}m_{(t)ab}$ as follows:

$$
\begin{aligned}
{}^{r+1}m_{(t)ab} &= \frac{\sum_{\boldsymbol{x},\boldsymbol{s}\backslash t,t+1} {}^{r+1}n(x_1, x_2, \ldots, x_T; s_1, s_2, \ldots, s_{t-1}, a, b, s_{t+2}, \ldots, s_T)}{\sum_{\boldsymbol{x},\boldsymbol{s}\backslash t} {}^{r+1}n(x_1, x_2, \ldots, x_T; s_1, s_2, \ldots s_{t-1}, a, s_{t+1}, \ldots, s_T)}, \\
&\quad a, b = 1, 2, \ldots, A; t = 1, 2, \ldots, T-1.
\end{aligned} \tag{5.20}
$$

Finally, the parameter estimation procedure for model (5.5) with (5.6) is constructed. Let ${}^r\boldsymbol{\varphi} = \left(({}^r v_a), ({}^r m_{(t)ab}), ({}^r p_{(t)ax}) \right)$ be the estimates at the rth iteration in the M-step; and let ${}^{r+1}D_{complete} = \left\{ {}^{r+1}n(x_1, x_2, \ldots, x_T; s_1, s_2, \ldots, s_T) \right\}$ be the conditional expectations of the complete data $D_{complete}$ at the $r+1$th iteration in the E-step. The model is an extended version of model (5.3) with (5.4), so the EM algorithm is presented as follows:

(i) E-step

$$
\begin{aligned}
&{}^{r+1}n(x_1, x_2, \ldots, x_T; s_1, s_2, \ldots, s_T) \\
&\quad = n(x_1, x_2, \ldots, x_T) \frac{{}^r v_{s_1} \prod_{t=1}^{T-1} {}^r m_{(t)s_t s_{t+1}} \prod_{t=1}^{T} {}^r p_{(t)s_t x_t}}{\sum_{\boldsymbol{s}} {}^r v_{s_1} \prod_{t=1}^{T-1} {}^r m_{(t)s_t s_{t+1}} \prod_{t=1}^{T} {}^r p_{(t)s_t x_t}}.
\end{aligned} \tag{5.21}
$$

(ii) M-step

Estimates ${}^{r+1}v_a$ and ${}^{r+1}m_{(t)ab}$ are given by (5.14) and (5.20), respectively. Estimates ${}^{r+1}p_{(t)s_t x_t}$ are calculated as follows:

$$
\begin{aligned}
{}^{r+1}p_{(t)ab} &= \frac{\sum_{,\boldsymbol{x}\backslash t,\boldsymbol{s}\backslash t} {}^{r+1}n(x_1, x_2, \ldots, x_{t-1}, b, x_{t+1}, \ldots x_T; s_1, s_2, \ldots, s_{t-1}, a, s_{t+1}, \ldots, s_T)}{\sum_{,\boldsymbol{x},,\boldsymbol{s}\backslash t} {}^{r+1}n(x_1, x_2, \ldots, x_{t-1}, x_t, x_{t+1}, \ldots x_T; s_1, s_2, \ldots, s_{t-1}, a, s_{t+1}, \ldots, s_T)}, \\
&\quad a = 1, 2, \ldots, A; b = 1, 2, \ldots, J; t = 1, 2, \ldots, T.
\end{aligned} \tag{5.22}
$$

5.4 A Property of the ML Estimation Procedure via the EM Algorithm

The parameter estimation procedures in the previous section have the following properties.

Theorem 5.2 In the parameter estimation procedures (5.11)–(5.16) for the time-homogeneous latent Markov chain model (5.1) with (5.2), if some of the initial trial values ${}^{0}v_a$, ${}^{0}m_{ab}$, and ${}^{0}p_{ab}$ are set to extreme values 0 or 1, then, the iterative values are automatically fixed to the values in the algorithm.

Proof Let us set ${}^{0}p_{ab} = 0$ for given a and b. From (5.12), we have

$$ {}^{1}n(x_1, x_2, \ldots, x_{t-1}, b, x_{t+1}, \ldots x_T; s_1, s_2, \ldots, s_{t-1}, a, s_{t+1}, \ldots, s_T) = 0. $$

By using the above values, formula (5.16) derives ${}^{1}p_{ab} = 0$. Hence, inductively it follows that

$$ {}^{t}p_{ab} = 0, t = 1, 2, \ldots. $$

In the other cases,

$$ {}^{0}p_{ab} = 1; {}^{0}v_a = 0, 1; {}^{0}m_{ab} = 0, 1, $$

the theorem holds true. ■

For the parameter estimation procedure for model (5.3) with (5.4), a similar property can be proven. In this sense, the estimation procedures based on the EM algorithm are convenient for the constraints.

By use of prior information about the phenomena concerned, some of the model parameters may be fixed to the extreme values, 0 s or 1 s. Especially, it may be significant to place the constraints on transition matrices. For example, let us consider a state transition diagram for latent state space $\Omega_{latent} = \{1, 2, 3\}$ shown in Fig. 5.2. From this figure, in model (5.1) with (5.2), the transition matrix is made as follows:

$$ M = \begin{pmatrix} 1 & 0 & 0 \\ 0 & m_{22} & m_{23} \\ m_{31} & m_{32} & m_{33} \end{pmatrix}. $$

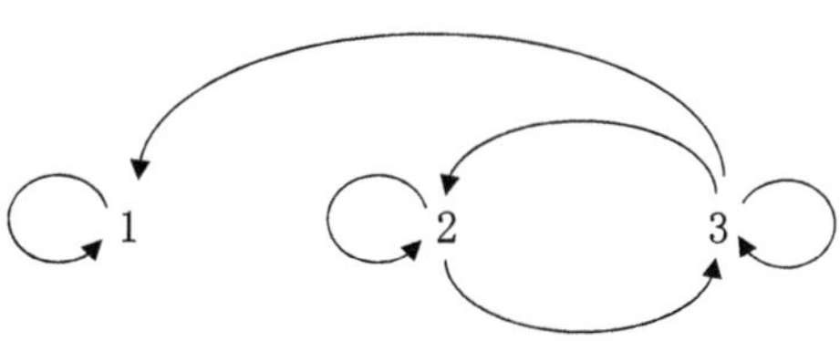

Fig. 5.2 A state transition path diagram

The above model can be estimated via the procedure mentioned in the previous section by setting

$$^{0}m_{12} = {}^{0}m_{13} = {}^{0}m_{21} = 0.$$

From Theorem 5.2, the above values are held fixed through iterations, that is,

$$^{r}m_{12} = {}^{r}m_{13} = {}^{r}m_{21} = 0, r = 1, 2, \ldots.$$

In model (5.1) with (5.2), if we set

$$A = J, {}^{0}p_{ab} = 0, a \neq b,$$

the estimation procedure derives the parameter estimates for the time-homogeneous Markov chain model. In a general model (5.5) with (5.6), if we set

$$^{0}m_{(t)ab} = 0, a \neq b,$$

and formally identify the states X_t at time points t as item responses, the EM algorithm for the model derives the ML estimates of the usual latent class model.

5.5 Numerical Example I

Table 5.1 shows the data [15] obtained by observing the changes in the configuration of interpersonal relationships in a group of 25 pupils at three time points: September, November, and January. They were asked "with whom would you like to sit?", and considering the state of each pair of pupils, the state concerned is one of the following three states: mutual choice, one-way choice, and indifference that are coded as "2", "1", and "0", respectively. The observation was carried out three times, two-monthly. In this case, the latent Markov chain model (5.1) or (5.5) can be used. First, the data are analyzed by use of the latent Markov chain models and the Markov chain models with three latent classes (states). The results of the analysis are shown in Table 5.2, and the estimated parameters of Markov models are illustrated in Tables 5.3 and 5.4. The latent Markov chain models fit the data set well according to the log likelihood test statistic G^2,

$$G^2 = 2\sum_{x} n(x_1, x_2, \ldots, x_T)\log\frac{n(x_1, x_2, \ldots, x_T)}{\hat{p}(x_1, x_2, \ldots, x_T)},$$

where $\hat{p}(x_1, x_2, \ldots, x_T)$ are the ML estimates of $p(x_1, x_2, \ldots, x_T)$. The above statistic is asymptotically χ^2- distributed with degrees of freedom "the number

Table 5.1 Data set I

Response pattern	Observed frequency	Response pattern	Observed frequency	Response pattern	Observed frequency
S N J		S N J		S N J	
0 0 0	197	1 0 0	15	2 0 0	3
0 0 1	20	1 0 1	6	2 0 1	0
0 0 2	0	1 0 2	0	2 0 2	0
0 1 0	12	1 1 0	6	2 1 0	0
0 1 1	9	1 1 1	9	2 1 1	3
0 1 2	1	1 1 2	4	2 1 2	2
0 2 0	0	1 2 0	3	2 2 0	3
0 2 1	0	1 2 1	0	2 2 1	0
0 2 2	1	1 2 2	2	2 2 2	4

* S: September; N: November; J: January
Source Katz and Proctor [15]

Table 5.2 Results of the analysis of data set I with Markov models

Model	G^2	df	P-val.	AIC
Time-homogeneous Markov chain model	27.903	18	0.064	100.369
Time-homogeneous latent Markov chain model	18.641	16	0.288	95.106
Non-homogeneous Markov chain model	17.713	14	0.220	98.179
Non-homogeneous latent Markov chain model	12.565	12	0.401	97.030

of manifest response patterns $(x_1, x_2, \ldots, x_T)$ minus 1" minus "the number of estimated parameters". Based on AIC [1], the time-homogeneous latent Markov chain model is selected as the most suitable one for explaining the data set, where

$$AIC = -2 \times (the\, maximum\, log\, likelihood) \\ + 2 \times (the\, number\, of\, estimated\, parameters).$$

5.6 Numerical Example II

Table 5.5 shows an artificial data set, and binary variables $X_{ti}, t = 1, 2$ imply manifest variables for the same items $i = 1, 2, 3$, where variables $X_{ti}, i = 1, 2, 3$ are indicators of latent variables $S_t, t = 1, 2$, assuming the three questions are asked to the same individuals at two time points. All variables are binary, so the states are denoted as 1 and 0. The data sets are made in order to demonstrate the estimation

Table 5.3 Estimated parameters in the time-homogeneous Markov models

Markov chaim model				Latent Markov chain model			
	Initial distribution				*Initial dstribution*		
	State				Latent state		
	0	1	2		0	1	2
	0.800	0.150	0.050		0.816	0.122	0.062
	Transition matrix				*Transition matrix*		
	State				Latent state		
State	0	1	2	Latent State	0	1	2
0	0.898	0.100	0.002	0	0.957	0.043	0.000
1	0.426	0.440	0.134	1	0.116	0.743	0.141
2	0.321	0.179	0.500	2	0.279	0.101	0.621
	Latent response probability				*Latent response probability*		
Latent State	0	1	2	Latent State	0	1	2
0	1*	0*	0*	0	0.952	0.043	0.000
1	0*	1*	0*	1	0.208	0.792	0.000
2	0*	0*	1*	2	0.000	0.135	0.815

* The numbers 0 and 1 are fixed

Table 5.4 Estimated parameters in the non-time-homogeneous Markov models

	Markov chaim model				Latent Markov chain model			
		Initial distribution				*Initial distribution*		
		State				Latent state		
		0	1	2		0	1	2
		0.800	0.150	0.050		0.823	0.127	0.050
		Transition matrix				*Transition matrix*		
		State				Latent state		
	State	0	1	2	Latent State	0	1	2
Sept.	0	0.904	0.092	0.004	0	0.959	0.040	0.001
to	1	0.467	0.422	0.111	1	0.225	0.625	0.146
Nov.	2	0.200	0.333	0.467	2	0.191	0.343	0.467
Nov.	0	0.892	0.108	0.000	0	0.942	0.058	0.000
to	1	0.391	0.457	0.152	1	0.138	0.681	0.181
Jan.	2	0.462	0.000	0.538	2	0.462	0.000	0.538
		Latent response probability				*Latent response probability*		
	Latent State	0	1	2	Latent State	0	1	2
	0	1*	0*	0*	0	0.953	0.047	0.000
	1	0*	1*	0*	1	0.121	0.879	0.000
	2	0*	0*	1*	2	0.000	0.000	1.000

Table 5.5 An artificial data set for a longitudinal observation

Time point							Time point						
1			2				1			2			
X_{11}	X_{12}	X_{13}	X_{21}	X_{22}	X_{23}	Freq	X_{11}	X_{12}	X_{13}	X_{21}	X_{22}	X_{23}	Freq
0	0	0	0	0	0	14	0	0	0	0	0	1	27
1	0	0	0	0	0	13	1	0	0	0	0	1	12
0	1	0	0	0	0	5	0	1	0	0	0	1	6
1	1	0	0	0	0	26	1	1	0	0	0	1	30
0	0	1	0	0	0	25	0	0	1	0	0	1	51
1	0	1	0	0	0	5	1	0	1	0	0	1	12
0	1	1	0	0	0	2	0	1	1	0	0	1	2
1	1	1	0	0	0	4	1	1	1	0	0	1	0
0	0	0	1	0	0	16	0	0	0	1	0	1	4
1	0	0	1	0	0	25	1	0	0	1	0	1	8
0	1	0	1	0	0	11	0	1	0	1	0	1	1
1	1	0	1	0	0	75	1	1	0	1	0	1	14
0	0	1	1	0	0	22	0	0	1	1	0	1	17
1	0	1	1	0	0	6	1	0	1	1	0	1	3
0	1	1	1	0	0	3	0	1	1	1	0	1	0
1	1	1	1	0	0	3	1	1	1	1	0	1	0
0	0	0	0	1	0	9	0	0	0	0	1	1	2
1	0	0	0	1	0	15	1	0	0	0	1	1	3
0	1	0	0	1	0	6	0	1	0	0	1	1	0
1	1	0	0	1	0	31	1	1	0	0	1	1	6
0	0	1	0	1	0	12	0	0	1	0	1	1	4
1	0	1	0	1	0	3	1	0	1	0	1	1	0
0	1	1	0	1	0	1	0	1	1	0	1	1	0
1	1	1	0	1	0	0	1	1	1	0	1	1	0
0	0	0	1	1	0	38	0	0	0	1	1	1	0
1	0	0	1	1	0	86	1	0	0	1	1	1	4
0	1	0	1	1	0	33	0	1	0	1	1	1	2
1	1	0	1	1	0	191	1	1	0	1	1	1	16
0	0	1	1	1	0	51	0	0	1	1	1	1	5
1	0	1	1	1	0	21	1	0	1	1	1	1	3
0	1	1	1	1	0	6	0	1	1	1	1	1	0
1	1	1	1	1	0	10	1	1	1	1	1	1	0

Table 5.6 The estimated parameters from data set II

	Initial distribution		Transition matrix		
	Latent state			Latent state	
	1	0	Latent state	1	0
	0.644	0.356	1	0.854	0.146
			0	0.492	0.508
Response probability for manifest variable					
Latent State	X_{t1}	X_{t2}	X_{t3}		
1	0.858	0.736	0.054		
0	0.197	0.055	0.681		

Log likelihood ration statistic $G^2 = 46.19, df = 54, P = 0.766.$

procedure for model (5.9) with two latent classes $A = 2$ and the number of observation time points $T = 2$. Since the ML estimation procedure via the EM algorithm can be constructed as in Sect. 5.3, the details are left for readers. The results of the parameter estimation are given in Table 5.6. According to the transition matrix, latent state "1" may be interpreted as a conservative one, and latent state 2 a less conservative one. In effect, the latent state distribution at the second time point is calculated by

$$\left(0.644\ 0.356\right)\begin{pmatrix}0.854 & 0.146\\ 0.492 & 0.508\end{pmatrix} = \left(0.725\ 0.275\right),$$

and it implies that the individuals with the first latent state are increased. If necessary, the distributions of $S_t, t \geq 3$ are calculated by

$$\left(0.644\ 0.356\right)\begin{pmatrix}0.854 & 0.146\\ 0.492 & 0.508\end{pmatrix}^{t-1}, t = 3, 4, \ldots .$$

5.7 A Latent Markov Chain Model with Missing Manifest Observations

Before constructing a more general model, a data set given in Bye and Schechter [8] is discussed. Table 5.7 illustrates the data from Social Security Administration services, and the individuals were assessed as severe or not severe with respect to the extent of work limitations. The observations were made in 1971, 1972, and 1974, where response "severe" is represented as "1" and "not severe" "0". The interval between 1972 and 1974 is two years and that between 1971 and 1972 is one year, that is, the time interval between 1972 and 1974 is twice as long as that between 1971 and 1972. When applying the Markov model to Data Set II, it is valid to assume that the observation in 1973 was missed, though the changes of latent states took place, that is. the transitions of manifest and latent states are $X_1 \to X_2 \to X_3$ and $S_1 \to S_2 \to U \to S_3$, respectively. Thus, the joint state transition can be expressed by

$$(X_1, S_1) \to (X_2, S_2) \to U \to (X_3, S_3).$$

In order to analyze the data set, a more general model was proposed by Bye and Schechter [8]. By using the notations in the previous section, for the time-homogeneous latent Markov chain model, the accounting equations are given by

$$p(x_1, x_2, x_3) = \sum_{\boldsymbol{S}} \sum_{u=1}^{A} v_{s_1} m_{s_1 s_2} m_{s_2 u} m_{u s_3} \prod_{t=1}^{3} p_{s_t x_t}, \tag{5.23}$$

where $\boldsymbol{s} = (s_1, s_2, s_3)$. The parameter estimation procedure via the EM algorithm (5.11–5.16) for the time-homogeneous Markov chain model is modified as follows:

(i) E-step

$$\begin{aligned} &{}^{r+1}n(x_1, x_2, x_3; s_1, s_2, s_3; u) \\ &= n(x_1, x_2, x_4) \frac{{}^{r}v_{s_1}{}^{r}m_{s_1 s_2}{}^{r}m_{s_2 u}{}^{r}m_{u s_3}{}^{r}p_{s_1 x_1}{}^{r}p_{s_2 x_2}{}^{r}p_{s_3 x_3}}{\sum_{\boldsymbol{s}} \sum_{u=1}^{A} {}^{r}v_{s_1}{}^{r}m_{s_1 s_2}{}^{r}m_{s_2 u}{}^{r}m_{u s_3}{}^{r}p_{s_1 x_1}{}^{r}p_{s_2 x_2}{}^{r}p_{s_3 x_3}}. \end{aligned} \tag{5.24}$$

(ii) M-step

$${}^{r+1}v_a = \frac{\sum_{\boldsymbol{x},\boldsymbol{s}\backslash 1} \sum_{u=1}^{A} {}^{r+1}n(x_1, x_2, x_3; a, s_2, s_3; u)}{N}, a = 1, 2, \ldots, A; \tag{5.25}$$

$$\begin{aligned} &{}^{r+1}p_{ab} \\ &= \frac{\sum_{t=1}^{T-1} \sum_{,\boldsymbol{x}\backslash t,\boldsymbol{s}\backslash t} \sum_{u=1}^{A} {}^{r+1}n(x_1, x_2, \ldots, x_{t-1}, b, x_{t+1}, \ldots x_T; s_1, s_2, \ldots, s_{t-1}, a, s_{t+1}, \ldots, s_T; u)}{\sum_{t=1}^{T-1} \sum_{,\boldsymbol{x},,\boldsymbol{s}\backslash t,} \sum_{u=1}^{A} {}^{r+1}n(x_1, x_2, \ldots, x_{t-1}, x_t, x_{t+1}, \ldots x_T; s_1, s_2, \ldots, s_{t-1}, a, s_{t+1}, \ldots, s_T; u)} \\ &T = 3, a = 1, 2, \ldots, A; b = 1, 2, \ldots, J, \end{aligned} \tag{5.26}$$

$${}^{r+1}m_{ab} = \frac{D_{ab} + E_{ab} + F_{ab}}{\sum_{b=1}^{A}(D_{ab} + E_{ab} + F_{ab})}, \tag{5.27}$$

where

$$D_{ab} = \sum_{\boldsymbol{x},\boldsymbol{s}\backslash 1,2} \sum_{u=1}^{A} {}^{r+1}n(x_1, x_2, x_3; a, b, s_3 : u),$$

Table 5.7 Data set II

Response pattern	Observed frequency	Response pattern	Observed frequency
$t_1t_2t_3$		$t_1t_2t_3$	
0 0 0	145	1 0 0	39
0 0 1	47	1 0 1	34
0 1 0	18	1 1 0	41
0 1 1	45	1 1 1	219

$^*t_1$: 1971; t_2 : 1972; t_3 : 1974
Source Bye and Schechter [8]

$$E_{ab} = \sum_{\boldsymbol{x},\boldsymbol{s}\backslash 2}{}^{r+1} n(x_1, x_2, x_3; s_1, a, s_3; b),$$

$$F_{ab} = \sum_{\boldsymbol{x},\boldsymbol{s}\backslash 3}{}^{r+1} n(x_1, x_2, x_3; s_1, s_2, b; a).$$

For the data set, model (5.23) is used for $A = 2$, $J = 2$ and the ML estimates of the parameters are obtained with the above procedure. Testing the goodness-of-fit of the model to the data set, we have $G^2 = 4.754$, $df = 2$, and $P = 0.093$ and it implies the goodness-of-fit of the model to the data set is fair. The estimates of the parameters are given in Table 5.8.

Remark 5.3 Bye and Schechter [8] uses the Newton method to obtain the ML estimates of the parameters in (5.23) for $A = J = 2$. In order to keep the following constraints

Table 5.8 The ML estimates of the parameters in model (5.23)

Latent Markov chain model		
	Initial distribution	
	Latent state	
	0	1
	0.430	0.570
	Transition matrix	
	Latent state	
Latent state	0	1
0	0.914	0.086
1	0.047	0.953
	Latent Response probability	
Latent state	0	1
0	0.897	0.103
1	0.097	0.903

$$0 < v_a < 1;\ 0 < m_{a1} < 1,\ ;\ 0 < p_{ai} < 1,\ a = 1, 2,\ i = 0.1, \tag{5.28}$$

the following parameter transformation was employed:

$$v_1 = \frac{1}{1 + \exp(\beta)},\ v_2 = \frac{\exp(\beta)}{1 + \exp(\beta)};$$

$$m_{a1} = \frac{1}{1 + \exp(\delta_a)},\ m_{a2} = \frac{\exp(\delta_a)}{1 + \exp(\delta_a)},\ a = 1, 2,$$

$$p_{a0} = \frac{1}{1 + \exp(\varepsilon_a)},\ p_{a1} = \frac{\exp(\varepsilon_a)}{1 + \exp(\varepsilon_a)},\ a = 1, 2.$$

However, the above transformation is model-specific, that is, (5.23) for $A = J = 2$, and a general transformation for multicategories $A > 2$ and/or $J > 2$ makes the ML estimation to be complicated. In contrast to it, the EM method such as (5.22–5.27) can be used easily, and the estimates always satisfy constraints

$$0 < v_a < 1; 0 < m_{ab} < 1, ; 0 < p_{aj} < 1, a = 1, 2, \ldots, A; j = 1, 2, \ldots, J. \tag{5.29}$$

In this respect, the EM method is superior to the Newton–Raphson method with the above parameter transformation.

5.8 A General Version of the Latent Markov Chain Model with Missing Manifest Observations

The time-homogeneous latent Markov chain model mentioned in the previous section [8] is extended to a general one. For observed time points t_i, $i = 1, 2, \ldots, T$, manifest responses X_i, $i = 1, 2, \ldots, T$ are observed, and the responses depend on latent states S_i at time points, where time points t_i are assumed integer, such that

$$t_1 < t_2 < \ldots < t_T \tag{5.30}$$

If interval $t_{i+1} - t_i > 1$, there are $t_{i+1} - t_i - 1$ time points (integers) between t_i and t_{i+1}, and at the time points the latent states u_{ij}, $j = 1, 2, \ldots, t_{i+1} - t_i - 1$ are changed as follows:

$$(s_i \rightarrow) u_{i1} \rightarrow u_{i2} \rightarrow \cdots \rightarrow u_{ih_i} (\rightarrow s_{i+1}), h_i = t_{i+1} - t_i - 1, \tag{5.31}$$

whereas the manifest states are not observed at the time points when the above sequences of latent states take place, for example, in Data Set II, it is assumed u_{21}

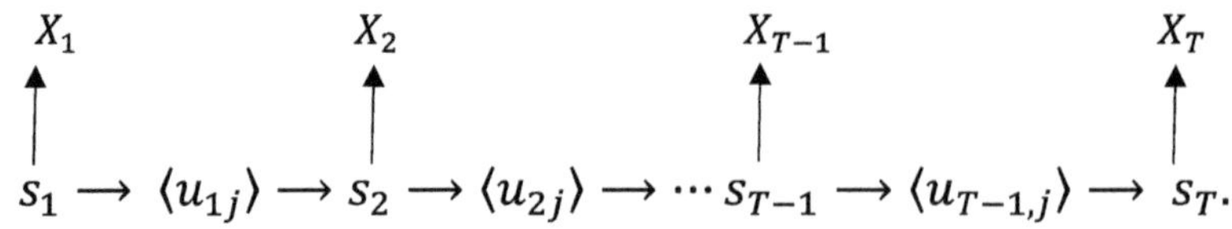

Fig. 5.3 Path diagram of latent Markov chaim model (5.33)

would occur at time point 1973. The above chain (5.31) is denoted as $\langle u_{ij} \rangle, i = 1, 2, \ldots, T - 1$ for short. Then, the changes of latent states are expressed as

$$s_1 \to \langle u_{1j} \rangle \to s_2 \to \langle u_{2j} \rangle \to \cdots \to \langle u_{T-1,j} \rangle \to s_T. \tag{5.32}$$

The manifest variables X_t are observed with latent state S_t and the responses depend on the latent states at time points $t, t = 1, 2, \ldots, T$. This model is depicted in Fig. 5.3. It is assumed that sequence (5.32) with (5.31) is distributed according to a time-homogeneous Markov chain with transition matrix $\left(m_{ij}\right)$. Let $p\left(x_1, x_2, \ldots, x_T; s_1, , s_2, , \ldots, , s_T; \langle u_{1j} \rangle, \langle u_{2j} \rangle, \ldots, \langle u_{T-1,j} \rangle\right)$ be the joint probabilities of manifest responses $(x_1, x_2, \ldots, x_T)$ and latent state transition (5.32). Then, we have

$$\begin{aligned} & p\left(x_1, x_2, \ldots, x_T; s_1, , s_2, , \ldots, , s_T; \langle u_{1j} \rangle, \langle u_{2j} \rangle, \ldots, \langle u_{T-1,j} \rangle\right) \\ & = \sum_{S} v_{s_1} \prod_{i=1}^{T} p_{s_i x_i} \times \prod_{t=1}^{T-1} \left(m_{s_t u_{t1}} m_{u_{t,h_i} s_{t+1}} \prod_{j=1}^{h_t - 1} m_{u_{tj} u_{t,i+1}} \right), \end{aligned} \tag{5.33}$$

where

$$m_{s_t u_{t1}} m_{u_{t,h_i} s_{t+1}} \prod_{j=1}^{h_t - 1} m_{u_{tj} u_{t,i+1}} = \begin{cases} m_{s_t u_{t1}} m_{u_{t1} s_{t+1}} & (h_t = 1) \\ m_{s_t s_{t+1}} & (h_t = 0) \end{cases}. \tag{5.34}$$

In repeated measurements, the time units are various, for example, day, week, month, and year. Although the observations are planned to make at regular intervals, there may be cases where the practices are not carried out. On such occasions., the above model may be feasible to apply to the cases. The ML estimation procedure via the EM algorithm can be constructed by extending (5.24) to (5.27).

5.9 The Latent Markov Process Model

Human behavior or responses takes place continuously in time; however, our observation is made in discrete time points, for example, daily, weekly, monthly, and so on. In this section, the change of latent states in Ω_{latent} is assumed to occur in a continuous time. Before constructing the model, the Markov process model is briefly reviewed [13]. It is assumed that an individual in a population takes decisions to change states in time interval $(0, t)$ according to a Poisson distribution with mean λt, where $\lambda > 0$, and that state changes are taken place with a Markov chain with transition matrix $\boldsymbol{Q} = (q_{ij})$. Let $t_i, i = 1, 2, \ldots$ be the decision time points that are taken place, and let $S(t)$ be a latent state at time point t. Then, given the time points $t_i, i = 1, 2, \ldots,$ the following sequence is distributed according to the Markov chain with transition matrix $\boldsymbol{Q}$:

$$S(t_1) \to S(t_2) \to \cdots \to S(t_n) \to \ldots.$$

The process is depicted in Fig. 5.4. Let $\boldsymbol{P}(t) = (p_{ij}(t))$ be the transition matrix of Markov process $S(t)$ on state space Ω_{latent} at time point t. Then, we have

$$\boldsymbol{M}(t) = \sum_{n=0}^{\infty} e^{-\lambda t} \frac{(\lambda t)^n}{n!} \boldsymbol{Q}^n, \tag{5.35}$$

where for $J \times J$ identity matrix $\boldsymbol{E}$, we set $\boldsymbol{Q}^0 = \boldsymbol{E}$. By differentiating the above matrix function with respect to time t, it follows that

$$\begin{aligned}
\frac{d}{dt}\boldsymbol{M}(t) &= -\lambda \sum_{n=0}^{\infty} e^{-\lambda t} \frac{(\lambda t)^n}{n!} \boldsymbol{Q}^n + n\lambda \sum_{n=0}^{\infty} e^{-\lambda t} \frac{(\lambda t)^{n-1}}{n!} \boldsymbol{Q}^n \\
&= -\lambda \boldsymbol{M}(t) + \lambda \boldsymbol{Q} \sum_{n=0}^{\infty} e^{-\lambda t} \frac{(\lambda t)^n}{n!} \boldsymbol{Q}^n \\
&= -\lambda \boldsymbol{M}(t) + \lambda \boldsymbol{Q}\boldsymbol{M}(t) = \lambda(\boldsymbol{Q} - \boldsymbol{E})\boldsymbol{M}(t).
\end{aligned}$$

Setting $\boldsymbol{R} = \lambda(\boldsymbol{Q} - \boldsymbol{E})$, we have the following differential equation:

$$\frac{d}{dt}\boldsymbol{M}(t) = \boldsymbol{R}\boldsymbol{M}(t).$$

From the above equation, given the initial condition $\boldsymbol{P}(t) = \boldsymbol{E}$, we get

$$\boldsymbol{M}(t) = \exp(\boldsymbol{R}t), \tag{5.36}$$

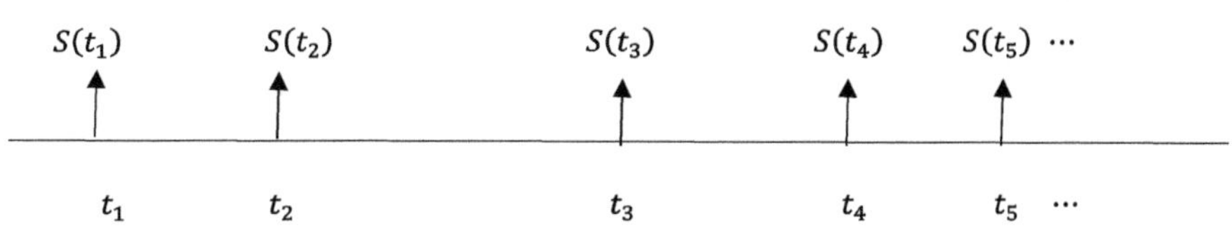

Fig. 5.4 Decision time points and latent state transitions

where for square matrix $\boldsymbol{B}$, we set

$$\exp(\boldsymbol{B}) \equiv \sum_{n=0}^{\infty} \frac{1}{n!} \boldsymbol{B}^n,$$

where $\frac{1}{0!}\boldsymbol{B}^0 \equiv \boldsymbol{E}$. In (5.36), matrix $\boldsymbol{R} = \left(r_{ij}\right)$ is called a generator matrix and, from the definition, the following constraints hold:

$$r_{ii} \leq 0; r_{ij} \geq 0, i \neq j; \sum_{j=1}^{J} r_{ij} = 0. \tag{5.37}$$

From (5.35), for $t, u > 0$, we also have

$$\boldsymbol{M}(t+u) = \exp(\boldsymbol{R}(t+u)) = \exp(t\boldsymbol{R})\exp(u\boldsymbol{R}) = \boldsymbol{M}(t)\boldsymbol{M}(u). \tag{5.38}$$

Especially, for integer k and $t = k\Delta t$, from (5.38), it follows that

$$\boldsymbol{M}(k\Delta t) = \boldsymbol{M}(\Delta t)^k. \tag{5.39}$$

From the above equation, if we observe a change of states at every time interval Δt (Fig. 5.5), the following sequence is the Markov chain with transition matrix $\boldsymbol{M}(\Delta t)$:

$$S(\Delta t) \to S(2\Delta t) \to \cdots \to S(k\Delta t) \to \ldots.$$

Considering the above basic discussion, the latent Markov process model is constructed. Let $t_i, i = 1, 2, \ldots, K$ be time points to observe manifest states $X(t_i)$ on finite state space $\Omega_{manifest}$; let $S(t)$ be the latent Markov process with generator matrix $\boldsymbol{R}$ with constraints (5.37); let $\boldsymbol{M}(t) = \left(m_{(t)ij}\right)$ be the transition matrix at time point t; and let p_{sx} be the probabilities of $X(t_i) = s$, given $S(t_i) = s$. For simplicity of the notation, given the time points, we use the following notation:

$$X_i = X(t_i);\ S_i = S(t_i), i = 1, 2, \ldots, K.$$

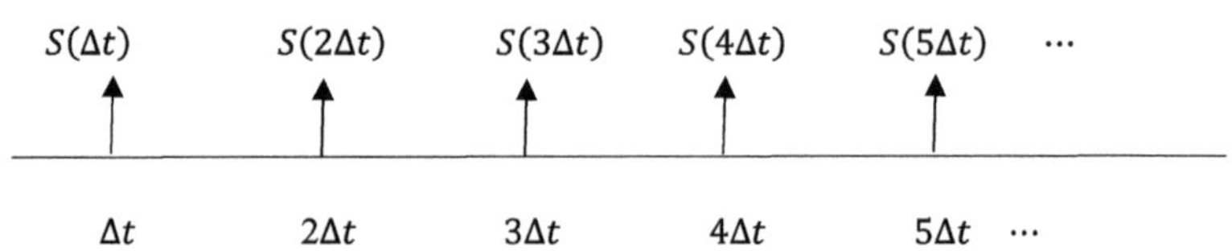

Fig. 5.5 Markov process due to observations at equal time intervals

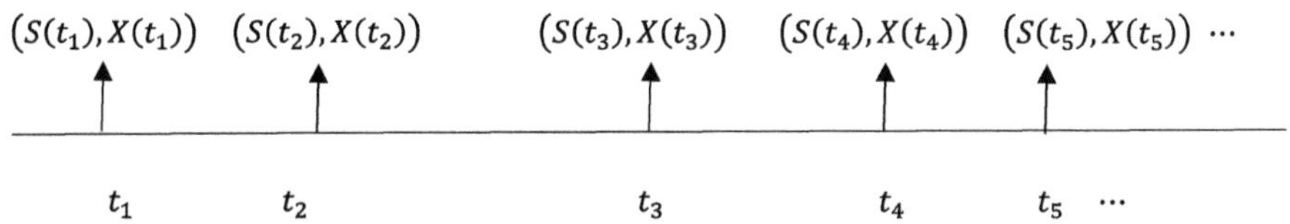

Fig. 5.6 The latent Markov process due to observations at any time interval

Then, by using similar notations as for the latent Markov chain modes, the following accounting equations can be obtained:

$$p(x_1, x_2, \ldots, x_T) = \sum_S v_{s_1} \prod_{i=1}^{K-1} m_{(t_{i+1}-t_i)s_t s_{t+1}} \prod_{j=1}^{K} p_{s_t x_t}. \tag{5.40}$$

The above model $(X(t), S(t))$ is called the latent Markov process model in the present chapter. The process is illustrated in Fig. 5.6.

In order to estimate the model parameters v_a, r_{ij}, and p_{ab}, the estimation procedure in the previous section may be used, because it is complicated to make a procedure for getting the estimates of r_{ij} directly, that is, parameters r_{ij} are elements of generator matrix $\boldsymbol{R}$ in (5.36). Usually, repeated observations of state changes are carried out in the intervals with time units, for example, daily, weekly, monthly, and so on, as in Table 5.7, so such time points can be viewed as integers. From this, for transition matrix $\boldsymbol{M}(t) = \left(m_{(t)ij}\right)$, we have

$$\boldsymbol{M}(t_{i+1} - t_i) = \exp(\boldsymbol{R}(t_{i+1} - t_i)) = \boldsymbol{M}(1)^{t_{i+1}-t_i}. \tag{5.41}$$

In order to simplify the notation, setting $\boldsymbol{M}(1) \equiv \boldsymbol{M} = (m_{ab})$, the same treatment of the model as in Sect. 5.7 may be conducted. If the estimates of transition probabilities m_{ab} can be available, we will estimate generator matrix $\boldsymbol{R}$ by the formal inversion:

$$\boldsymbol{R} = \log \boldsymbol{M} = \sum_{n=1}^{\infty} \frac{(-1)^{n-1}}{n} (\boldsymbol{M} - \boldsymbol{E})^n, \tag{5.42}$$

where $\boldsymbol{E}$ is the identity matrix, whereas it is an important question whether there exists a Markov process with the transition matrix [2]. The problem is called that of *Embeddability*. Singer and Spilerman [17] gave the following necessary condition for obtaining the generator matrix.

Theorem 5.3 If the eigenvalues of transition matrix $\boldsymbol{M}$ are positive and distinct, any solution of $\boldsymbol{M} = \exp(\boldsymbol{R})$ is unique. ■

Remark 5.4 In the above theorem, for $A \times A$ matrix $\boldsymbol{M}$, let $\rho_a, a = 1, 2, \ldots, A$ be the positive and distinct eigenvalues. Then, there exists a non-singular matrix $\boldsymbol{C}$, and the transition matrix is expressed by

$$M = CDC^{-1},$$

where

$$D = \begin{pmatrix} \rho_1 & 0 & \cdots 0 \\ 0 & \rho_2 & \cdots 0 \\ \vdots & \vdots & \ldots \vdots \\ 0 & 0 & \cdots \rho_A \end{pmatrix}.$$

From this, we can get

$$R = \log CDC^{-1} = C \begin{pmatrix} \log\rho_1 & 0 & \cdots 0 \\ 0 & \log\rho_2 & \cdots 0 \\ \vdots & \vdots & \ldots \vdots \\ 0 & 0 & \cdots \log\rho_A \end{pmatrix} C^{-1}. \quad (5.43)$$

The result is the same as calculated by (5.42). ■

For the number of latent states $A = 2$, the following transition matrix is considered:

$$M = \begin{pmatrix} m_{11} & m_{12} \\ m_{21} & m_{22} \end{pmatrix}. \quad (5.44)$$

Under the condition $m_{a1} + m_{a2} = 1, a = 1, 2$, the characteristic equation is given by

$$(x - 1)(x - m_{11} - m_{22} + 1) = 0.$$

From the equation, we have two eigenvalues 1 and $m_{11} + m_{22} - 1$. If

$$(2 >)m_{11} + m_{22} > 1, \quad (5.45)$$

from (5.42), the matrix equation

$$M = \exp(R)$$

is solved via (5.43). For transition matrix (5.44), we have

$$M = \begin{pmatrix} 1 & m_{12} \\ 1 & -m_{21} \end{pmatrix} \begin{pmatrix} 1 & 0 \\ 0 & m_{11} + m_{22} - 1 \end{pmatrix} \begin{pmatrix} 1 & m_{12} \\ 1 & -m_{21} \end{pmatrix}^{-1}.$$

It follows that

$$\boldsymbol{R}\left(=\begin{pmatrix} r_{11} & r_{12} \\ r_{21} & r_{22} \end{pmatrix}\right)=\begin{pmatrix} 1 & m_{12} \\ 1 & -m_{21} \end{pmatrix}\begin{pmatrix} 0 & 0 \\ 0 & \log(m_{11}+m_{22}-1) \end{pmatrix}\begin{pmatrix} 1 & m_{12} \\ 1 & -m_{21} \end{pmatrix}^{-1}$$
$$=\begin{pmatrix} m_{12}\log(m_{11}+m_{22}-1) & -m_{12}\log(m_{11}+m_{22}-1) \\ -m_{21}\log(m_{11}+m_{22}-1) & m_{21}\log(m_{11}+m_{22}-1) \end{pmatrix}. \tag{5.46}$$

From (5.45), since $1 > m_{11} + m_{22} - 1 > 0$, we see

$$m_{12}\log(m_{11}+m_{22}-1) < 0, m_{21}\log(m_{11}+m_{22}-1) < 0,$$

and the condition for the generator matrix (5.37) is met for $A = 2$. Thus, the transition matrix (5.44) is embeddable under the condition (5.45). Applying the above discussion to Table 5.8, since

$$\widehat{m}_{11} + \widehat{m}_{22} = 0.957 + 0.743 = 1.700 > 1,$$

condition (5.45) is satisfied by the estimated transition matrix. From Theorem 5.3, there exists a unique generator matrix in equation $\boldsymbol{M} = \exp(\boldsymbol{R})$ and then, by using (5.43), we have

$$\widehat{\boldsymbol{R}} = \begin{pmatrix} -0.085 & 0.085 \\ 0.046 & -0.046 \end{pmatrix}.$$

According to the above generator matrix, the transition matrix can be calculated at any time point by (5.36), for example, we have

$$\boldsymbol{M}(1.5) = \begin{pmatrix} 0.875 & 0.125 \\ 0.068 & 0.932 \end{pmatrix}, \boldsymbol{M}(2.5) = \begin{pmatrix} 0.806 & 0.194 \\ 0.106 & 0.894 \end{pmatrix},$$

$$\boldsymbol{M}(3.5) = \begin{pmatrix} 0.746 & 0.254 \\ 0.139 & 0.861 \end{pmatrix}, \boldsymbol{M}(4.5) = \begin{pmatrix} 0.694 & 0.306 \\ 0.167 & 0.833 \end{pmatrix}, \ldots,$$

$$\boldsymbol{M}(\infty) = \begin{pmatrix} 0.353 & 0.647 \\ 0.353 & 0.647 \end{pmatrix}.$$

Next, for the following transition matrix with three latent states:

$$\boldsymbol{M} = \begin{pmatrix} m_{11} & m_{12} & m_{13} \\ m_{21} & m_{22} & m_{23} \\ m_{31} & m_{32} & m_{33} \end{pmatrix},$$

the characteristic function is calculated as

$$
\begin{aligned}
\det(\boldsymbol{M} - x\boldsymbol{E}) &= \det\begin{pmatrix} m_{11} - x & m_{12} & m_{13} \\ m_{21} & m_{22} - x & m_{23} \\ m_{31} & m_{32} & m_{33} - x \end{pmatrix} \\
&= (1 - x)\det\begin{pmatrix} 1 & m_{12} & m_{13} \\ 1 & m_{22} - x & m_{23} \\ 1 & m_{32} & m_{33} - x \end{pmatrix} \\
&= (1 - x)\det\begin{pmatrix} 1 & 0 & 0 \\ 1 & m_{22} - m_{12} - x & m_{23} - m_{13} \\ 1 & m_{32} - m_{12} & m_{33} - m_{13} - x \end{pmatrix} \\
&= (1 - x)\det\begin{pmatrix} m_{22} - m_{12} - x & m_{23} - m_{13} \\ m_{32} - m_{12} & m_{33} - m_{13} - x \end{pmatrix} = 0.
\end{aligned}
$$

Setting

$$
\begin{pmatrix} g_{11} & g_{12} \\ g_{21} & g_{22} \end{pmatrix} = \begin{pmatrix} m_{22} - m_{12} & m_{23} - m_{13} \\ m_{32} - m_{12} & m_{33} - m_{13} \end{pmatrix}, \tag{5.47}
$$

if

$$
\begin{cases} (g_{11} - 1)(g_{22} - 1) - g_{12}g_{21} \neq 0, \\ g_{11}g_{22} - g_{12}g_{21} > 0, \\ g_{11} + g_{22} > 0, \end{cases} \tag{5.48}
$$

from Theorem 5.3, the above matrix has a unique matrix $\boldsymbol{R}$ such that $\boldsymbol{M} = \exp(\boldsymbol{R})$.

Remark 5.5 Condition (5.48) does not always imply the matrix obtained with (5.42) is the generator matrix of a Markov process. ■

The above discussion is applied to the estimated transition matrices of the time-homogeneous Markov models in Table 5.3. For the Markov chain model, the estimated transition matrix is

$$
\widehat{\boldsymbol{M}} = \begin{pmatrix} 0.898 & 0.100 & 0.002 \\ 0.426 & 0.440 & 0.134 \\ 0.321 & 0.179 & 0.500 \end{pmatrix}.
$$

From (5.47), we have

$$
\begin{pmatrix} g_{11} & g_{12} \\ g_{21} & g_{22} \end{pmatrix} = \begin{pmatrix} 0.340 & 0.132 \\ 0.079 & 0.498 \end{pmatrix},
$$

and from (5.48) the sufficient condition for getting the unique solution $\boldsymbol{R}$ of $\boldsymbol{M} = \exp(\boldsymbol{R})$ is checked as follows:

$$\begin{cases} (g_{11}-1)(g_{22}-1) - g_{12}g_{21} = 0.321, \\ g_{11}g_{22} - g_{12}g_{21} = 0.159, \\ g_{11} + g_{22} = 0.838. \end{cases}$$

From this, the three conditions in (5.48) are met. In effect, we can get the generator matrix by (5.43) as follows:

$$\begin{aligned} \widehat{\boldsymbol{R}} &= \begin{pmatrix} 1 & 0.145 & 0.138 \\ 1 & -0.491 & -0.848 \\ 1 & -0.859 & 0.512 \end{pmatrix} \begin{pmatrix} \log 1 & 0 & 0 \\ 0 & \log 0.548 & 0 \\ 0 & 0 & \log 290 \end{pmatrix} \begin{pmatrix} 1 & 0.145 & 0.138 \\ 1 & -0.491 & -0.848 \\ 1 & -0.859 & 0.512 \end{pmatrix}^{-1} \\ &= \begin{pmatrix} -0.148 & 0.166 & -0.018 \\ 0.641 & -0.949 & 0.307 \\ 0.382 & 0.361 & -0.743 \end{pmatrix}. \end{aligned}$$

However, the above estimate of the generator matrix is improper, because the condition in (5.37) is not met, that is, $\hat{r}_{13} = -0.018 < 0$. For the latent Markov chain model in Table 5.3, the transition matrix is estimated as

$$\widehat{\boldsymbol{M}} = \begin{pmatrix} 0.957 & 0.043 & 0.000 \\ 0.116 & 0.743 & 0.141 \\ 0.279 & 0.101 & 0.621 \end{pmatrix}.$$

The eigenvalues of the above transition matrix are distinct and positive. Through a similar discussion above, we have the estimate of the generator matrix of the latent Markov chain model as

$$\widehat{\boldsymbol{R}} = \begin{pmatrix} -0.046 & 0.051 & -0.005 \\ 0.104 & -0.314 & 0.210 \\ 0.353 & 0.140 & -0.493 \end{pmatrix}.$$

However, this case also gives an improper estimate of the generator. Although the ML estimation procedure for the continuous-time mover-stayer model is complicated, Cook et al. [9] proposed a generalized version of the model, and gave an ML estimation procedure for the model.

5.10 Discussion

This chapter has considered the latent Markov chain models for explaining changes of latent states in time. The ML estimation procedures of the models have been constructed via the EM algorithm, and the methods are demonstrated by using numerical examples. As in Chap. 2, the latent states in the latent Markov chain model are treated parallelly, so in this sense, the analysis can be viewed as latent class cluster analysis as well, though the latent Markov analysis is a natural extension of the latent class analysis. In confirmatory contexts as discussed in Sect. 5.4, as shown in Theorem 5.2, the ML estimation procedures are flexible to handle the constraints for setting some of the model parameters to the extreme values 0 s or 1 s. As in Chaps. 3 and 4, it is important to make the latent Markov models structured for measurement of latent states in an ability or trait, in which logit models are effective to express the effects of latent states [3]. For the structure of latent state transition, logit models with the effects of covariates have been applied to the initial distribution and the transition matrices in the latent Markov chain model [19], and the extensions have also been studied by Bartolucci et al. [4], Bartolucci and Pennoni [5], and Bartolucci and Farcomeni [3]. Such approaches may link to path analysis with generalized linear models ([10–12], Chap. 6), and further studies for extending latent Markov approaches will be expected.

References

1. Akaike, H. (1987). Factor analysis and AIC. *Psychometrika, 52*, 317–332.
2. Bartholomew, D. J. (1983). Some recent development in social statistics. *International Statistical Review, 51*, 1–9.
3. Bartolucci, F., & Farcomeni, A. (2009). A multivariate extension of the dynamic logit model for longitudinal data based on a latent Markov heterogeneity structure. *Journal of the American Statistical Association, 104*, 816–831.
4. Bartolucci, F., Pennoni, F., & Francis, B. (2007). A latent Markov model for detecting patterns of criminal activity. *Journal of the Royal Statistical Society, A, 170*, 115–132.
5. Bartolucci, F., Farcomeni, A., & Pennoni, F. (2014). Latent Markov models: A review of a general framework for the analysis of longitudinal data with covariates. *TEST, 23*, 433–465.
6. Baum, L., & Petrie, T. (1966). Statistical inference for probabilistic functions of finite state Markov chains. *Annals of Mathematical Statistics, 37*, 1554–1563.
7. Baum, L., Petrie, T., Soules, G., & Weiss, N. (1970). A maximization technique occurring in the statistical analysis of probabilistic functions of Markov chains. *Annals of Mathematical Statistics, 41*, 164–171.
8. Bye, B. V., & Schechter, E. S. (1986). A latent Markov model approach to the estimation of response error in multiway panel data. *Journal of the American Statistical Association, 51*, 702–704.
9. Cook, R. J., Kalbfleisch, J. D., & Yi, G. Y. (2002). A generalized mover-stayer model for panel data. *Biostatistics, 3*, 407–420.
10. Eshima, N. (2020). *Statistical data analysis and entropy*. Springer.
11. Eshima, N., Tabata, M., Borroni, C. G., & Kano, Y. (2018). An entropy-based approach to path analysis of structural generalized linear models: A basic approach. *Entropy, 17*, 5117–5132.

12. Eshima, N., Tabata, M., & Zhi, G. (2001). Path analysis with logistic regression models: Effect analysis of fully recursive causal systems of categorical variables. *Journal of the Japan Statistical Society, 31*, 1–14.
13. Hatori, H., & Mori, T. (1993). *Finite Markov Chains, Faifukan*: Tokyo (in Japanese).
14. Lazarsfeld, P. F., & Henry, N. M. (1968). *Latent structure analysis*. Houghton Mifflin.
15. Katz, L., & Proctor, C. (1959). The concept of configuration of interpersonal relation in a group as a time-dependent stochastic process. *Psychometrika, 24*, 317–327.
16. Singer, B., & Spilerman, S. (1975). Identifying structural parameters of social processes using fragmentary data. *Bulletin of International Statistical Institute, 46*, 681–697.
17. Singer, B., & Spilerman, S. (1976). The representation of social processes by Markov models. *American Journal of Sociology, 82*, 1–54.
18. Singer, B., & Spilerman, S. (1977). Fitting stochastic models to longitudinal survey data—some examples in the social sciences. *Bulletin of International Statistical Institute, 47*, 283–300.
19. Vermunt, J. K., Langeheine, R., & Bockenholt, U. (1999). Discrete-time discrete-state latent Markov models with time-constant and time-varying covariates. *Journal of Educational and Behavioral Statistics, 24*, 179–207.

Chapter 6
The Mixed Latent Markov Chain Model

6.1 Introduction

As a model that explains time-dependent human behavior, the Markov chain has been applied in various scientific fields [1, 2, 6, 7]. When employing the model for describing human behavior, it may be usually assumed that every individual in a population changes his or her states at any observational time point according to the same low of probability, as an approximation; however, there are cases where the population is not homogeneous, and it makes the analysis of human response processes to be complicated [9]. In order to overcome the heterogeneity of the population, it is valid to consider the population is divided into subpopulations that depend on the Markov chains of their own. In order to analyze the heterogeneous population, Blumen et al. [8] proposed the mover-stayer model, in which the population is divided into two subpopulations of, what we call, "movers" and "stayers". The movers change their states according to a Markov chain and the stayers do not change their states from the initial observed time points. Human behavior, which is observed repeatedly, is more complicated than the mover-stayer model. An extended version of the mover-stayer model is the mixed Markov chain model that was introduced first by Poulsen, C. S. in 1982 in his Ph. D. dissertation, though the work was not officially published [18, 19]. Eshima et al. [12, 13], Bye & Schechter [10], Van de Pol & de Leeuw [20], and Poulsen [17] also discussed similar topics. Van de Pol [18] proposed the mixed latent Markov chain model as an extension of the mixed Markov chain model. Figure 1 shows the relation of the above Markov models, in which the arrows indicate the natural directions of extension.

Following Chap. 5, this chapter provides a discussion of dynamic latent structure analysis within a framework of the latent Markov chain model [11]. In Sect. 6.2, dynamic latent structure models depicted in Fig. 6.1 are briefly reviewed, and the equivalence of the latent Markov chain model and the mixed latent Markov chain model is shown. Section 6.3 discusses the ML estimation procedure for the models via the EM algorithm in relation to that for the latent Markov chain model. In Sect. 6.4,

N. Eshima, *An Introduction to Latent Class Analysis*, Behaviormetrics: Quantitative Approaches to Human Behavior 14,
https://doi.org/10.1007/978-981-19-0972-6_6

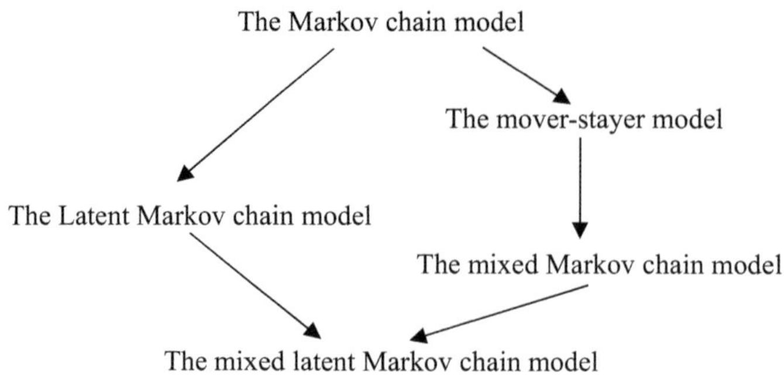

Fig. 6.1 Relation of Markov models. *The directions expressed by the arrows imply the extensions of the related models

a numerical example is given to demonstrate the method for the ML estimation. Finally, Sect. 6.5 briefly reviews advanced studies for the mixed Markov modeling to link to further research.

6.2 Dynamic Latent Class Models

In the present section, the mover-stayer model, the mixed Markov chain model, and the mixed latent Markov chain model are reviewed.

(i) The mover-stayer model

Let $\Omega = \{1, 2, \ldots, A\}$ be the manifest state space of the Markov chain. It is assumed that a population is divided into two types of individuals, movers and stayers, and the proportions are set as λ_1 and λ_2, respectively, where

$$\lambda_1 + \lambda_2 = 1.$$

Let v_a be probabilities that take the initial states a of the mover and let w_a be those of the stayer at the first time point, $a = 1, 2, \ldots, A$. Then,

$$\sum_{a=1}^{A} v_a = \sum_{a=1}^{A} w_a = 1.$$

Suppose that the mover changes the states according to a time-homogeneous Markov chain. Let m_{ab} be the transition probabilities from state a at time point t to state b at time point $t + 1$, and let $p(x_1, x_2, \ldots, x_T)$ be the probabilities of manifest state transition, $x_1 \to x_2 \to \cdots \to x_T$. Then, we have the following accounting equations:

$$p(x_1, x_2, \ldots x_T) = \lambda_1 \sum_{s} v_{x_1} \prod_{t=1}^{T-1} m_{x_t x_{t+1}} + \lambda_2 \sum_{s} w_{x_1} \prod_{t=1}^{T-1} \delta_{x_t x_{t+1}}, \tag{6.1}$$

where the above summation is made over all response patterns $\boldsymbol{s} = (s_1, s_2, \ldots, s_T)$ and

$$\delta_{x_t x_{t+1}} = \begin{cases} 1 & x_t = x_{t+1} \\ 0 & x_t \neq x_{t+1} \end{cases}.$$

Remark 6.1 When the Markov chain in (6.1) is time-dependent, the transition probabilities for movers $m_{x_t x_{t+1}}$ are substituted for $m_{(t)x_t x_{t+1}}, t = 1, 2, \ldots, T-1$.

(ii) The mixed Markov chain model

It is assumed that a population is divided into K subpopulations that depend on time-homogeneous Markov chains of their own. Let ψ_k be the proportions of subpopulations $k = 1, 2, \ldots, K$; $v_{ka}, a = 1, 2, \ldots, A$ be the initial state distributions of Markov chain k, and let $m_{kx_t x_{t+1}}$ be transition probabilities from manifest state x_t at time point t to x_{t+1} at time point $t+1$. The manifest state space is $\Omega = \{1, 2, \ldots, A\}$. Then, the accounting equations are given by

$$p(x_1, x_2, \ldots, x_T) = \sum_{k=1}^{K} \psi_k v_{kx_1} \prod_{t=1}^{T-1} m_{kx_t x_{t+1}}, \tag{6.2}$$

where

$$\sum_{k=1}^{K} \psi_k = 1, \sum_{a=1}^{A} v_{ka} = 1, \sum_{b=1}^{A} m_{kab} = 1.$$

For $K = 2$, setting

$$m_{2ab} = \begin{cases} 1 & a = b \\ 0 & a \neq b \end{cases},$$

model (6.2) expresses the mover-stayer model (6.1).

(iii) The mixed latent Markov chain model

Let latent variables $S_t, t = 1, 2, \ldots, T$ be Markov chains on $\Psi = \{1, 2, \ldots, C\}$ in subpopulation $k = 1, 2, \ldots, K$; let $X_t, t = 1, 2, \ldots, T$ be manifest variables on state space $\Omega = \{1, 2, \ldots, A\}$, and let $p_{ks_{kt} x_t}$ be the conditional probability of $X_t = x_t$ given latent state $S_t = s_t$ in subpopulation k at time point t. Then, the response probabilities are expressed as follows:

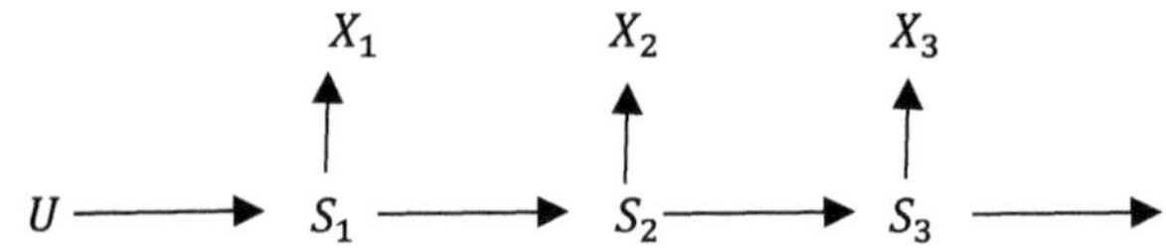

Fig. 6.2 Path diagram of the mixed latent Markov chain model (6.3)

$$p(x_1, x_2, \ldots, x_T) = \sum_{k=1}^{K} \psi_k \sum_{s} v_{kx_1} p_{ks_1x_1} \prod_{t=1}^{T-1} \left(p_{ks_{t+1}x_{t+1}} m_{ks_ts_{t+1}}\right), \tag{6.3}$$

where

$$\sum_{k=1}^{K} \psi_k = 1, \sum_{b=1}^{B} v_{kb} = 1, \sum_{a=1}^{A} p_{kba} = 1, \sum_{c=1}^{B} m_{kbc} = 1. \tag{6.4}$$

For $A = B$ and $p_{kba} = \delta_{ab}$, where δ_{ab} is the Kronecker delta, (6.3) expresses (6.2), and setting $K = 1$, model (6.3) becomes the latent Markov chain model (5.1).

Remark 6.2 When manifest response probabilities and transition ones are dependent on observed time points t, p_{kab} and m_{kab} are replaced by $p_{(t)kab}$ and $m_{(t)kab}$, respectively. ■

Let $U = k \in \{1, 2, \ldots, K\}$ be a categorical latent variable that expresses subpopulations depending on the latent Markov chains with the initial state distributions (v_{ka}) and the transition matrices (m_{kab}). Then, it means that the conditional distribution of sequence $S_1 \to S_2 \to \cdots \to S_T$ given $U = k$ is a Markov chain with the initial distribution $\left(v_{(k)a}\right)$ and transition matrix $\left(m_{(k)ab}\right)$, and the path diagram of variables U, $\{S_t\}$, and $\{X_t\}$ is illustrated in Fig. 6.2. Although a natural direction in the extension of latent structure models is illustrated in Fig. 6.1, we have the following theorem.

Theorem 6.1 *The latent Markov chain model and the mixed latent Markov chain model are equivalent.*

Proof In latent Markov chain model (5.1) with $B = CK$ latent states, let latent state space $\Psi = \{1, 2, \ldots, B\}$ of latent variables S_t be divided into K subspaces as $\Psi_k = \{C(k-1)+1, C(k-1)+2, \ldots, Ck\}, k = 1, 2, \ldots, K$. If the subspaces are closed with respect to the state transition, the transition matrix of the latent Markov chain model is given as the following type:

$$\boldsymbol{M} = \begin{pmatrix} \boldsymbol{M}_1 & 0 & \cdots & 0 \\ 0 & \boldsymbol{M}_2 & \cdots & 0 \\ \vdots & \vdots & \ddots & \vdots \\ 0 & \cdots & \cdots & \boldsymbol{M}_K \end{pmatrix}, \tag{6.5}$$

where $\boldsymbol{M}_k, k = 1, 2, \ldots, K$ are $C \times C$ transition matrices with latent state spaces Ψ_k. In model (6.5), we have

$$\sum_{j=1}^{C} m_{C(k-1)+i, C(k-1)+j} = 1, i = 1, 2, \ldots, C; k = 1, 2, \ldots, K.$$

For the latent Markov chain model (5.1), setting

$$\lambda_k = \sum_{i=1}^{C} v_{C(k-1)+i}, v_{(k)c} = \frac{v_{C(k-1)+c}}{\lambda_k}, c = 1, 2, \ldots, C; k = 1, 2, \ldots, K, \quad (6.6)$$

it follows that

$$\sum_{c=1}^{C} v_{(k)c} = 1, k = 1, 2, \ldots, K.$$

and the latent Markov chain model expresses the mixed latent Markov chain model. This completes the theorem. ■

In Chap. 5, the equivalence of the latent class model and the latent Markov chain model is shown in Theorem 5.1, so the following result also holds true:

Theorem 6.2 *The mixed latent Markov chain model is equivalent to the latent class model.*

6.3 The ML Estimation of the Parameters of Dynamic Latent Class Models

In this section, the ML estimation procedures of latent class models via the EM algorithm are summarized. Although the methods can be directly made for the individual latent structure models as shown in the previous chapters, the ML estimation procedures are given through that for the latent Markov chain model in Chap. 5, based on Theorem 5.2.

(i) The mover-stayer model

Let the manifest and latent state space be set as $\Omega = \{1, 2, \ldots, A\}$ and $\Psi = \{1, 2, \ldots, 2A\}$, respectively, and let space Ψ be divided into the following two subspaces:

$$\Psi_1 = \{1, 2, \ldots, A\} \text{ and } \Psi_2 = \{A + 1, A + 2, \ldots, 2A\}.$$

If the initial trial values of estimates of the parameters in (5.15) and (5.16) in the ML estimation algorithm are put, respectively, as follows:

$$
{}^{0}m_{ab} = \begin{cases} 0 & (a \in \Psi_1, b \in \Psi_2; a \in \Psi_2, b \in \Psi_1) \\ 1 & (a = b \in \Psi_2) \\ 0 & (a \neq b \in \Psi_2) \end{cases}, \tag{6.7}
$$

and

$$
{}^{0}p_{ab} = \begin{cases} 1 & (a = b \, or \, a = b + A) \\ 0 & (otherwise) \end{cases} \tag{6.8}
$$

then, from (5.12), we have

$$
{}^{1}n(x_1, x_2, \ldots, x_T; s_1, s_2, \ldots, s_{t-1}, a, b, s_{t+2}, \ldots, s_T) = 0
$$
$$
(a \in \Psi_1, b \in \Psi_2; a \in \Psi_2, b \in \Psi_1; a \neq b \in \Psi_2), \tag{6.9}
$$

$$
{}^{1}n(x_1, x_2, \ldots, x_{t-1}, b, x_{t+1}, \ldots x_T; s_1, s_2, \ldots, s_{t-1}, a, s_{t+1}, \ldots, s_T) = 0
$$
$$
(a \neq b \text{ or } a \neq b + A). \tag{6.10}
$$

Putting (6.9) into (5.15), it follows that

$$
{}^{1}m_{ab} = \begin{cases} 0 & (a \in \Psi_1, b \in \Psi_2; a \in \Psi_2, b \in \Psi_1) \\ 1 & (a = b \in \Psi_2) \\ 0 & (a \neq b \in \Psi_2) \end{cases}.
$$

Similarly, by (6.10) we also have

$$
{}^{1}p_{ab} = \begin{cases} 1 & (a = b \, or \, a = b + A) \\ 0 & (otherwise) \end{cases}
$$

from (5.16). Inductively,

$$
{}^{r}p_{ab} = \begin{cases} 1 & (a = b \, or \, a = b + A) \\ 0 & (\text{otherwise}) \end{cases},
$$

$$
{}^{r}m_{ab} = \begin{cases} 0 & (a \in \Psi_1, b \in \Psi_2; a \in \Psi_2, b \in \Psi_1) \\ 1 & (a = b \in \Psi_2) \\ 0 & (a \neq b \in \Psi_2) \end{cases}, r = 1, 2, \ldots.
$$

Hence, by setting the initial trial values of the parameters as (6.7) and (6.8), the ML estimates of parameters in the mover-stayer model can be obtained via the EM algorithm for the latent Markov chain model in (5.11) through (5.16).

(ii) The mixed Markov chain model

For manifest state space $\Omega = \{1, 2, \ldots, A\}$, latent state space Ψ is divided into

$$\Psi_k = \{A(k-1)+1, A(k-1)+2, \ldots, Ak\}, k = 1, 2, \ldots, K,$$

where $\Psi = \bigcup_{k=1}^{K} \Psi_k$. If we set the initial trial values for (5.15) and (5.16) in the EM algorithm for the latent Markov chain model (5.1) as

$$^{0}m_{ab} = 0(a \in \Psi_k, b \in \Psi_l, k \neq l),$$

and

$$^{0}p_{ab} = \begin{cases} 1 \ (a = b + A(k-1), k = 1, 2, \ldots, K) \\ 0 \qquad\qquad (otherwise) \end{cases}$$

then, the algorithm estimates the mixed Markov chain model.

(iii) The mixed latent Markov chain model

By using the parameterization in Theorem 6.1 and identifying response probabilities p_{bx} in the latent Markov chain model (5.1) and p_{kbx} in the mixed latent Markov chain model (6.3) as

$$p_{kcx} = p_{c+C(k-1),x}, c = 1, 2, \ldots, C; x = 1, 2, \ldots, A; k = 1, 2, \ldots, K,$$

setting the initial trial values of the parameters as in (6.5), the algorithm for the latent Markov chain model in (5.11) through (5.16) makes the ML estimates of the parameters.

6.4 A Numerical Illustration

In order to demonstrate the above discussion, for the data set in Table 5.1, the mover-stayer model and the mixed latent Markov chain model are estimated via the EM algorithm for the latent Markov chain model in Chap. 5. The mover-stayer model is estimated with the method in the previous section. For the same patterns of initial trial values of the parameters (6.7) and (6.8), we have the ML estimates of the parameters in Table 6.1. According to the log likelihood test of goodness-of-fit to the data set,

Table 6.1 The estimated parameters in the mover-stayer model

Initial distribution							
	Mover (latent state)				Stayer (latent state)		
	1	2	3		4	5	6
	0.346	0.147	0.046		0.454	0.003	0.004
	Latent transition matrix						
	Latent state						
		1	2	3	4	5	6
Latent State	1	0.765	0.230	0.005	0*	0*	0*
	2	0.438	0.428	0.135	0*	0*	0*
	3	0.348	0.193	0.459	0*	0*	0*
	4	0*	0*	0*	1*	0*	0*
	5	0*	0*	0*	0*	1*	0*
	6	0*	0*	0*	0*	0*	1*
	Response probability						
	Manifest state						
			1	2	3		
	Latent State	1	1*	0*	0*		
		2	0*	1*	0*		
		3	0*	0*	1*		
		4	1*	0*	0*		
		5	0*	1*	0*		
		6	0*	0*	1*		
The log likelihood ratio statistic $G^2 = 23.385, df = 15, p = 0.076$							

The numbers with "*" imply the fixed values

$G^2 = 23.385, df = 15, p = 0.076$, the model is accepted at the significant level 0.05. Similarly, the ML estimation of the mixed latent Markov chain model is carried out, and the estimates of the parameters are illustrated in Table 6.2. The results do not provide a good fit for the data set. As illustrated in this section, the ML estimation of the latent structure models in Fig. 6.1 can be made practically by using the ML estimation procedure for the latent Markov chain model.

6.5 Discussion

The present chapter has treated a basic version of the mixed latent Markov chain model, in which it is assumed the response probabilities to test items at a time point and the transition probabilities depend only on the latent states at the time point. In this sense, the basic model gives an exploratory analysis similar to the latent

Table 6.2 The estimated parameters in the mixed latent Markov chain model

Initial distribution							
	Mover (latent state)				Stayer (latent state)		
	1	2	3		4	5	6
	0.215	0.124	0.033		0.611	0.000	0.017
	Latent transition matrix						
	Latent state						
		1	2	3	4	5	6
Latent State	1	0.727	0.265	0.008	0*	0*	0*
	2	0.368	0.491	0.141	0*	0*	0*
	3	0.385	0.000	0.615	0*	0*	0*
	4	0*	0*	0*	1.000	0.000	0.000
	5	0*	0*	0*	0.000	0.598	0.402
	6	0*	0*	0*	0.051	0.949	0.000
	Response probability						
	Manifest state						
			1	2	3		
	Latent State	1	0.968	0.032	0.000		
		2	0.000	1.000	0.000		
		3	0.000	0.000	1.000		
		4	0.968	0.032	0.000		
		5	0.000	1.000	0.000		
		6	0.000	0.000	1.000		

The log likelihood ratio statistic $G^2 = 8.809, df = 3, p = 0.032$.

The numbers with "*" imply the fixed values

class cluster analysis. As discussed here, the model is an extension of the latent structure models in Fig. 6.1 from a natural viewpoint; however, the mixed latent Markov chain model is equivalent to the latent Markov chain model, and also to the latent class model. The parameter estimation in the mixed latent Markov chain model via the EM algorithm can be carried out by using that for the latent Markov chain model as shown in Sect. 6.3, and the method has been demonstrated in Sect. 6.4. The estimation algorithm is convenient to handle the constraints for extreme values, setting a part of the response and transition probabilities as zeroes and ones, and the property is applied to the parameter estimation in the mixed latent Markov chain model. Applying the model to various research fields, there may be cases where the response probabilities to manifest variables and the transition probabilities are influenced by covariates and histories of latent state transitions, and for dealing with such cases, the mixed latent Markov chain models have been excellently developed in applications by Langeheine [16], Vermunt et al. [21], Bartolucci [3], Bartolucci & Farcomeni [4], Bartolucci et al. [5], and so on. Figure 6.3 illustrates the manifest

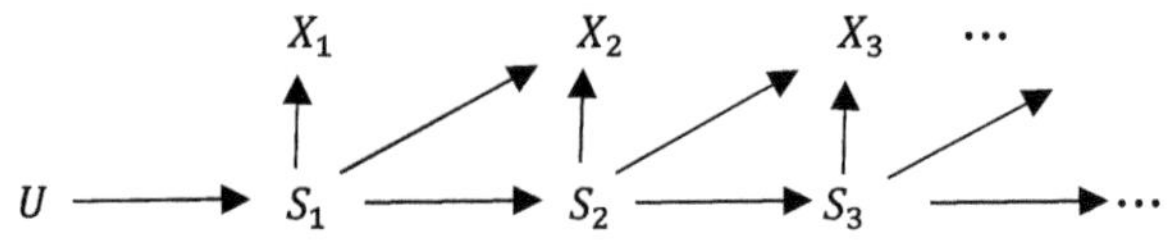

Fig. 6.3 A path diagram of the mixed latent Markov chain model with the effects of latent state histories on manifest variables

Fig. 6.4 A path diagram of the mixed latent Markov chain model with the effects of covariates and latent state histories on manifest variables

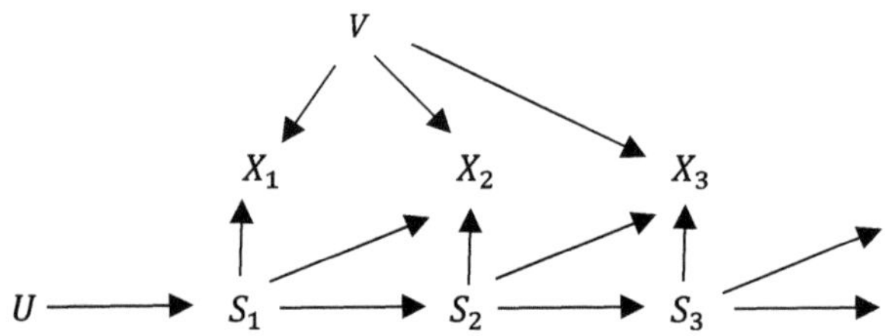

variables $X_t, t = 2, 3, ..$ depend on histories $S_{t-1} \rightarrow S_t$, and Fig. 6.4 shows covariate V influences manifest variables $X_t, t = 1, 2, \ldots$ in addition to latent state histories $S_{t-1} \rightarrow S_t$. In these cases, logit model approaches can be made as in Chaps. 3 and 4, for example, in Fig. 6.3, for binary latent variables S_t and binary manifest variables X_t, assuming there are no interaction effects of S_{t-1} and S_t on X_t, the following logit model can be made:

$$P(X_t = x_t | S_{t-1} = s_{t-1}, S_t = s_t) = \frac{\exp\left(x_t\alpha_{(t)} + x_t\beta_{(t)t-1}s_{t-1} + x_t\beta_{(t)t}s_t\right)}{1 + \exp\left(\alpha_{(t)} + \beta_{(t)t-1}s_{t-1} + \beta_{(t)t}s_t\right)}, t = 2, 3, \ldots,$$

where $\alpha_{(t)}$ and $\beta_{(t)t-1}$ are parameters. For polytomous variables, generalized logit models can be constructed by considering phenomena under study. Similarly, for Fig. 6.4, appropriate logit models can also be discussed. In such models, it may be useful to make path analysis of the system of variables. Further developments of latent Markov modeling in data analysis can be expected. In Chap. 7, an entropy-based approach to path analysis [14, 15] is applied to latent class models.

References

1. Andersen, E. B. (1977). *Discrete statistical models with social science application*. Amsterdam: North-Holland Publishing Co.
2. Bartholomew, D. J. (1983). Some recent development of social statistics. *International Statistical Review, 51*, 1–9.
3. Bartolucci, F. (2006). Likelihood inference for a class of latent Markov models under linear hypotheses on the transition probabilities. *Journal of the Royal Statistical Society, B, 68*, 155–178.
4. Bartolucci, F., & Farcomeni, A. (2009). A multivariate extension of the dynamic logit model for longitudinal data based on a latent Markov heterogeneity structure. *Journal of the American*

Statistical Association, 104, 816–831.

5. Bartolucci, F., Lupparelli, M., & Montanari, G. E. (2009). Latent Markov model for binary longituidinal data: An application to the performance evaluation of nursing homes. *Annals of Applied Statistics, 3*, 611–636.
6. Bartolucci, F., Farcomeni, A., & Pennoni, F. (2010). An overview of latent Markov models for longitudinal categorical data, arXiv:1003.2804 [math.ST].
7. Bartolucci, F., Farcomeni, A., & Pennoni, F. (2014). Latent Markov models: A review of a general framework for the analysis of longitudinal data with covariates. *TEST, 23*, 433–465.
8. Blumen, I., Kogan, M., & McCarthy, P. J. (1955). *The industry mobility of labor as a probability process*. Ithaca: Cornel University.
9. Bush, R. R., & Cohen, B. P. (1956). Book Review of. *Journal of the American Statistical Association, 51*, 702–704.
10. Bye, B. V., & Schechter, E. S. (1986). A latent Markov model approach to the estimation of response error in multiway panel data. *Journal of the American Statistical Association, 81*, 357–380.
11. Eshima, N. (1993). Dynamic latent structure analysis through the latent Markov chain model. *Behaviormetrika, 20*, 151–160.
12. Eshima, N, Asano, C, & Watanabe, M (1984). A time-dependent latent class analysis based on states-transition. In *Proceedings of the First China-Japan Symposium on Statistics*, pp. 62–66.
13. Eshima, N., Asano, C., & Watanabe, M. (1985). A time-dependent latent class analysis based on states-transition. *Sougo Rikougaku Kenkyuka Houkoku, 6*, 243–249. (in Japanese).
14. Eshima, N., Tabata, M., & Zhi, G. (2001). Path analysis with logistic regression models: Effect analysis of fully recursive causal systems of categorical variables. *Journal of Japan Statistical Society, 31*, 1–14.
15. Eshima, N., Tabata, M., Borroni, C. G., & Kano, Y. (2018). An entropy-based approach to path analysis of structural generalized linear models: a basic approach. *Entropy, 17*, 5117–5132.
16. Langeheine, R. (1988). Manifest and latent Markov chain models for categorical panel data. *Journal of Educational Statistics, 13*, 299–312.
17. Poulsen, C. S. (1990). Mixed Markov and latent Markov modelling applied to brand choice behaviour. *International Journal of Research in Marketing, 7*, 5–19.
18. Van de Pol, F. (1990). A unified framework for Markov modeling in discrete and discrete time. *Sociological Method and Research, 18*, 416–441.
19. Van de Pol, F., & Langeheine, R. (1990). Mixed Markov latent class models. *Sociological Methodology, 20*, 213–247.
20. Van de Pol, F., & de Leeuw, J. (1986). A latent Markov model to correct for measurement error. *Sociological Method and Research, 15*, 118–141.
21. Vermunt, J. K., Langeheine, R., & Böckenholt, U. (1999). Discrete-time discrete-state latent Markov models with time-constant and time-varying covariates. *Journal of Educational and Behavioral Statistics, 24*, 179–207.

Chapter 7
Path Analysis in Latent Class Models

7.1 Introduction

It is a useful approach to analyze causal relationships among variables in latent structure models. The relationships are considered in real data analysis based on observational methods of the variables and by using particular scientific theories, and according to it, causes and effects are hypothesized on sets of the variables before statistical analysis. In many scientific fields, for example, sociology, psychology, education, medicine, and so on, there are many cases where some of the observed variables are regarded as indicators of latent variables, so meaningful causal relationships have to be discussed not only for manifest variables but also for latent variables through path analysis methods [25]. Linear structural equation models (Jöreskog and Sörbom, 1996) [2] are significant approaches for path analysis of continuous variables, and the path analysis is easily carried out by using regression coefficients in linear structural equations that express path diagrams among manifest and latent variables. For categorical variables, Goodman's approach to path analysis with odds ratios [11, 12] made a great stimulus for developing path analysis of categorical variables, and Goodman [13] tried a direction of path analysis in latent class models, though the direct and indirect effects were not discussed. This approach was performed in the case where all variables concerned are binary, and the models used are called the multiple-indicator, multiple-cause models. Macready (1982) also used a latent class model with four latent classes, which represent learning patterns of two kinds of skill, to perform a causal analysis in explaining a learning structure. Similar models are also included in White and Clark [24], Owston [23], and Eshima et al. [7]. In path analysis, it is important how the total effects of parent variables on descendant ones are measured and how the total effects are divided into the direct and indirect effects, that is, the following additive decomposition is critical:

$$\text{The total effect} = \text{the direct effect} + \text{the indirect effect}. \tag{7.1}$$

N. Eshima, *An Introduction to Latent Class Analysis*, Behaviormetrics: Quantitative Approaches to Human Behavior 14,
https://doi.org/10.1007/978-981-19-0972-6_7

Eshima et al. [8] proposed path analysis for categorical variables in logit models by using log odds ratios, and the above decomposition was given. Kuha and Goldthorpe [16] also gave a path analysis method of categorical variables by using odds ratios, however, decomposition (7.1) approximately holds true. Following the approaches, an entropy-based method of path analysis for generalized linear models, which can make the effect decomposition shown in (7.1), was proposed by Eshima et al. [9].

The present chapter applies a method of path analysis in Eshima et al. [8, 9] to multiple-indicator, multiple-cause models and the latent Markov chain model. Section 7.2 discusses a multiple-indicator, multiple-cause model. In Sect. 7.3, the path analysis method is reviewed, and the effects of variables are calculated in some examples. Section 7.4 gives a numerical illustration to make a path analysis in the multiple-indicator, multiple-cause model. In Sect. 7.5, path analysis in the latent Markov chain model is considered, and in Sect. 7.6, a numerical example is presented to demonstrate the path analysis. Section 7.7 provides discussions and a further perspective of path analysis in latent class models.

7.2 A Multiple-Indicator, Multiple-Cause Model

Let binary variables $X_{ki}, i = 1,2,\ldots,I_k$ be the indicators of latent variables $S_k, k = 1,2,\ldots,K$, and let us suppose that the conditional probabilities of indicator variables X_{ki}, given $(S_1, S_2, \ldots, S_K) = (s_1, s_2, \ldots, s_K)$, depend only on $S_k = s_k$, that is, for $l \neq k$, latent variables S_l do not have the direct effects on $X_{ki}, i = 1,2,\ldots,I_k$. Let $v(s_1, s_2, \ldots, s_K)$ be the probability of $(S_1, S_2, \ldots, S_K) = (s_1, s_2, \ldots, s_K)$. Then, the latent class model is given as follows:

$$\begin{aligned}&\mathrm{P}(X_{ki} = x_{ki}, i = 1,2,\ldots,I_k; k = 1,2,\ldots K)\\&= \sum_s v(\boldsymbol{s}) \prod_{k=1}^{K} \prod_{i=1}^{I_k} P(X_{ki} = 1|S_k = s_k)^{x_{ki}} (1 - P(X_{ki} = 1|S_k = s_k))^{1-x_{ki}},\end{aligned} \tag{7.2}$$

where notation $\sum_s$ implies the summation over all latent variable patterns $\boldsymbol{s} = (s_1, s_2, \ldots, s_K)$. In this model, the following inequalities have to hold:

$$P(X_{ki} = 1|S_k = 0) < P(X_{ki} = 1|S_k = 1), i = 1,2,\ldots,I_k; k = 1,2,\ldots,K, \tag{7.3}$$

because $X_{ki}, i = 1,2,\ldots,I_k$ are the indicators of latent variables (states) $S_k, k = 1,2,\ldots,K$. The path diagram between S_k and $X_{ki}, i = 1,2,\ldots,I_k$ is illustrated in Fig. 7.1. The probabilities $P(X_{ki} = 1|S_k = 0)$ imply guessing (intrusion) errors and $1 - P(X_{ki} = 1|S_k = 1)$ forgetting (omission) ones. Although the parameters can be

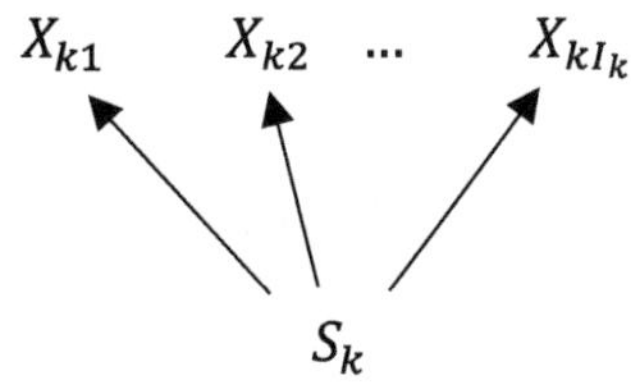

Fig. 7.1 Path diagram of manifest variables X_i and latent variable S_k

estimated according to the EM algorithm for the usual latent class model in Chap. 2, imposing the equality constraints, there may be cases that the estimated models are not identified with the hypothesized structures. Hence, in order to estimate the latent probabilities $P(X_{ki} = 1|S_k = s_k)$, it is better to formulate the latent probabilities as models in Chaps. 3 and 4, that is,

$$P(X_{ki} = 1|S_k = s_k) = \begin{cases} \frac{\exp(\alpha_{ki})}{1+\exp(\alpha_{ki})} & (s_k = 0) \\ \frac{\exp(\alpha_{ki}+\beta_{ki})}{1+\exp(\alpha_{ki}+\beta_{ki})} & (s_k = 1) \end{cases}, \tag{7.4}$$

where

$$\beta_{ki} = \exp(\gamma_{ki}), i = 1,2, \ldots, I_k; k = 1,2, \ldots, K. \tag{7.5}$$

The ML estimation procedure based on the EM algorithm can be constructed by a method similar to those in Chaps. 2 and 3. For $K = 2$, a path diagram of the model is illustrated in Fig. 7.2a. The total, direct and indirect effects have to be

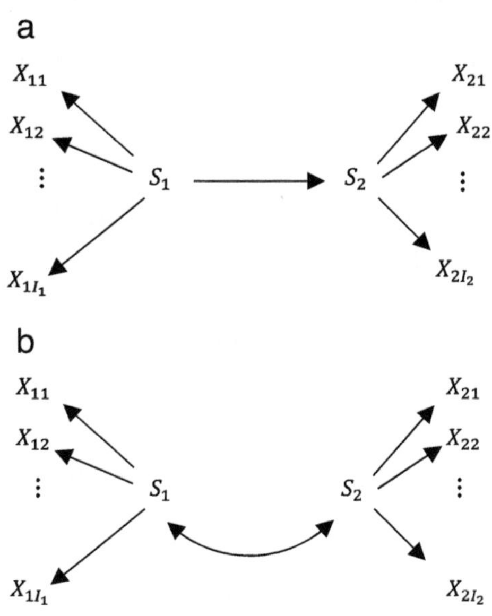

Fig. 7.2 **a** Path diagram of manifest variables $X_{ki}, i = 1,2, \ldots, I_k$ and latent variables $S_k, k = 1,2$, where S_1 is a parent variable of S_2. **b** Path diagram of manifest variables $X_{ki}, i = 1,2, \ldots, I_k$ and latent variables $S_k, k = 1,2$, where S_1 and S_2 have no causal order

calculated according to path diagrams. Let $f(s_1, s_2, x_{1i}, x_{2j})$ be the joint probability functions of variables $(S_1, S_2, X_{1i}, X_{2j})$, $g_1(s_1)$ the marginal probability function of S_1, $g_{12}(s_2|s_1)$ the conditional probability function of S_2 for given $S_1 = s_1$, $f_{1i}(x_{1i}|s_1)$ that of X_{1i} for given $S_1 = s_1$, and let $f_{2j}(x_{2j}|s_2)$ be that of X_{2j} for given $S_2 = s_2$. Then, from Fig. 7.2a, functions $f(s_1, s_2, x_{1i}, x_{2j})$ are decomposed as follows:

$$f(s_1, s_2, x_{1i}, x_{2j}) = g_1(s_1)g_{12}(s_2|s_1)f_{1i}(x_{1i}|s_1)f_{2j}(x_{2j}|s_2),$$
$$i = 1,2, \ldots, I_1;\ j = 1,2, \ldots, I_2. \tag{7.6}$$

Since $f_{ki}(x_{ki}|s_k) = P(X_{ki} = x_{ki}|S_k = s_k)$, for binary latent variables S_k, formulae in (7.4) are expressed as

$$f_{ki}(x_{ki}|s_k) = \frac{\exp(x_{ki}\alpha_{ki} + x_{ki}\beta_{ki}s_k)}{1 + \exp(\alpha_{ki} + \beta_{ki}s_k)}, i = 1,2, \ldots, I_k : k = 1,2. \tag{7.7}$$

Similarly, we have

$$g_{12}(s_2|s_1) = \frac{\exp(s_2\gamma_1 + s_2\delta_1 s_1)}{1 + \exp(\gamma_1 + \delta_1 s_1)}, s_1, s_2 \in \{0,1\}, \tag{7.8}$$

where δ_1 is a regression coefficient and γ_1 is an intercept parameter. From the above formulation, the path system in Fig. 7.2a is viewed as a recursive system of logit models. In Fig. 7.2a, S_1 is the parent variable of the other variables, and S_2 is the parent variable of manifest variables $X_{2i}, i = 1,2, \ldots, I_2$. If there is no causal order between latent variables S_1 and S_2, then, the path diagram is illustrated in Fig. 7.2b. In this model, the latent variables have to be treated parallelly. Before making a path analysis of the multiple-indicator, multiple-cause model (7.2), an entropy-based path analysis method for generalized linear model (GLM) systems [8, 9] is considered for logit models in the next section.

7.3 An Entropy-Based Path Analysis of Categorical Variables

For simplicity of the discussion, a system of variables Y and $U_i, i = 1,2,3$ shown in Fig. 7.3 is discussed. From the path diagram, the joint probability of the four variables is recursively decomposed as follows:

$$f(u_1, u_2, u_3, y) = f_1(u_1)f_2(u_2|u_1)f_3(u_3|u_1, u_2)f(y|u_1, u_2, u_3),$$

where functions $f_i(*|*), i = 1,2,3$ and $f(*|*)$ imply the conditional probability functions related to the variables. In Fig. 7.3, the relationship is expressed as follows:

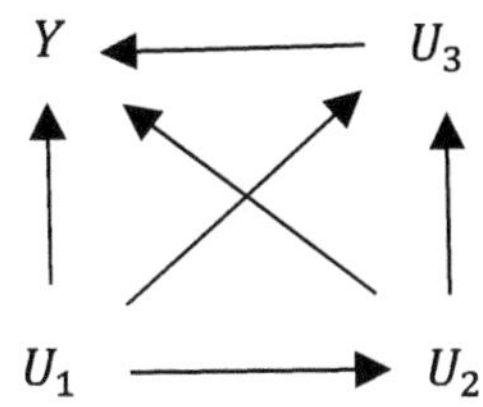

Fig. 7.3 A path diagram of manifest variable Y and latent variables $U_i, i = 1,2,3$

$$U_1 \prec U_2 \prec U_3 \prec Y.$$

For the following logit model without no interaction effects:

$$f(y|u_1, u_2, u_3) = \frac{\exp(y\alpha + y\beta_1 u_1 + y\beta_2 u_2 + y\beta_3 u_3)}{1 + \exp(\alpha + \beta_1 u_1 + \beta_2 u_2 + \beta_3 u_3)}, \tag{7.9}$$

the total, direct, and indirect effects of parent variables U_k on descendant variable Y are discussed. For baseline category $(U_1, U_2, U_3, Y) = \left(u_1^*, u_2^*, u_3^*, y^*\right)$, the total effect of $(U_1, U_2, U_3) = (u_1, u_2, u_3)$ on $Y = y$ can be defined by the following log odds ratio:

$$\begin{aligned}\log\frac{f(y|u_1, u_2, u_3) f\left(y^* u_1^*, u_2^*, u_3^*\right)}{f(y^*|u_1, u_2, u_3) f\left(y|u_1^*, u_2^*, u_3^*\right)} &= \left(\log f(y|u_1, u_2, u_3) - \log f\left(y^*|u_1, u_2, u_3\right)\right) \\ &\quad - \left(\log f\left(y|u_1^*, u_2^*, u_3^*\right) - f\left(y^*|u_1^*, u_2^*, u_3^*\right).\right) \\ &= \sum_{k=1}^{3}\left(y - y^*\right)\beta_k\left(u_k - u_k^*\right). \end{aligned} \tag{7.10}$$

The above log odds ratio implies the decrease of the uncertainty of response variable Y for a change of parent variables (U_1, U_2, U_3) from baseline $\left(u_1^*, u_2^*, u_3^*\right)$, that is, the amount of information on Y explained by the parent variables. Since the logit model in (7.9) has no interactive effects of the explanatory variables U_i, the log odds ratio is a bilinear form of $y - y^*$ and $u_k - u_k^*$ with respect to regression coefficients β_k. The baseline $\left(u_1^*, u_2^*, u_3^*, y^*\right)$ is formally substituted for the related means of the variables $(\mu_1, \mu_2, \mu_3, \nu)$ and the total effect of $(U_1, U_2, U_3) = (u_1, u_2, u_3)$ on $Y = y$ is defined by

$$\sum_{k=1}^{3}(y - \nu)\beta_k(u_k - \mu_k). \tag{7.11}$$

The total effect of (u_2, u_3) on $Y = y$ at $U_1 = u_1$ is defined by

$$\sum_{k=2}^{3}(y - \nu(u_1))\beta_k(u_k - \mu_k(u_1)), \tag{7.12}$$

where $\nu(u_1)$ and $\mu_k(u_1)$ are the conditional means of Y and $U_k, k = 2,3$ given $U_1 = u_1$, respectively. The above formula can be derived by formally setting $\left(u_1^*, u_2^*, u_3^*, y^*\right) = (u_1, \mu_2(u_1), \mu_3(u_1), \nu(u_1))$ in (7.10). Subtracting (7.11) from (7.12), it follows that

$$\begin{aligned} &\text{The total effect of } U_1 = u_1 \text{ on } Y = \text{y at } (U_2, U_3) = (u_2, u_3) \\ &\quad = (\text{the total effect of } (U_1, U_2, U_3) = (u_1, u_2, u_3) \text{ on } Y = y) \\ &\quad - (\text{the total effect of } (U_2, U_3) = (u_2, u_3) \text{ on } Y = y \text{ at } U_1 = u_1) \\ &\quad = \sum_{k=1}^{3} (y - \mu_Y)\beta_k(u_k - \mu_k) - \sum_{k=2}^{3} (y - \nu(u_1))\beta_k(u_k - \mu_k(u_1)). \end{aligned} \tag{7.13}$$

Putting $\left(u_1^*, u_2^*, u_3^*, y^*\right) = (\mu_1(u_2, u_3), u_2, u_3, \nu(u_2, u_3))$, where $\nu(u_2, u_3)$ and $\mu_1(u_2, u_3)$ are the conditional means of Y and U_1 given $(U_2, U_3) = (u_2, u_3)$, respectively, the direct effect of $U_1 = u_1$ on $Y = y$ at $(U_2, U_3) = (u_2, u_3)$ is defined by

$$(y - \nu(u_2, u_3))\beta_1(u_1 - \mu_1(u_2, u_3)). \tag{7.14}$$

From this, the indirect effect of $U_1 = u_1$ on $Y =$ y through $(U_2, U_3) = (u_2, u_3)$ is calculated by subtracting (7.14) from (7.13).

Remark 7.1 The indirect effect of $U_1 = u_1$ on $Y = y$ is defined by the total effect minus the direct effect as discussed above. Since the direct and the total effects of $U_1 = u_1$ can be interpreted as information, the indirect effect is also interpreted in information. ■

Second, the effects of $U_2 = u_2$ on $Y =$ y at $(U_1, U_3) = (u_1, u_3)$ are computed. The total effect of $U_3 = u_3$ on $Y =$ y at $(U_1, U_2) = (u_1, u_2)$ can be calculated by setting $\left(u_1^*, u_2^*, u_3^*, y^*\right) = (u_1, u_2, u_3(u_1, u_2), \nu(u_1, u_2))$ in (7.10), that is,

$$(y - \nu(u_1, u_2))\beta_3(u_3 - \mu_3(u_1, u_2)), \tag{7.15}$$

where $\nu(u_1, u_2)$ and $\mu_3(u_1, u_2)$ are the conditional means of Y and U_3 given $(U_1, U_2) = (u_1, u_2)$, respectively. From (7.12) and (7.15), the total effect of $U_2 = u_2$ on $Y =$ y at $(U_1, U_3) = (u_1, u_3)$ is calculated as follows:

$$\begin{aligned} &(\text{the total effect of } (U_2, U_3) = (u_2, u_3) \text{ on } Y = \text{y at } U_1 = u_1) \\ &\quad - (\text{The total effect of } U_3 = u_3 \text{ on } Y = y \text{ at } (U_1, U_2) = (u_1, u_2)) \\ &= \sum_{k=2}^{3} (y - \nu(u_1))\beta_k(u_k - \mu_k(u_1)) - (y - \nu(u_1, u_2))\beta_3(u_3 - \mu_3(u_1, u_2)) \end{aligned} \tag{7.16}$$

and for baseline $\left(u_1^*, u_2^*, u_3^*, y^*\right) = (u_1, \mu_2(u_1, u_3), u_3, \mu_Y(u_1, u_3))$ in (7.10), the direct effect $U_2 = u_2$ on $Y = y$ at $(U_1, U_3) = (u_1, u_3)$ is given by

$$(y - \nu(u_1, u_3))\beta_2(u_2 - \mu_2(u_1, u_3)), \tag{7.17}$$

where $\nu(u_1, u_3)$ and $\mu_2(u_1, u_3)$ are the conditional means of Y and U_2 given $(U_1, U_3) = (u_1, u_2)$, respectively. From the above calculation, we have the following additive decomposition of the total effect of $(U_1, U_2, U_3) = (u_1, u_2, u_3)$ on $Y = y$:

$$\begin{aligned}
&(\text{The total effect of } (U_1, U_2, U_3) = (u_1, u_2, u_3) \text{ on } Y) \\
&\quad = (\text{The total effect of } U_1 = u_1 \text{ on } Y = y \text{ at } (U_2, U_3) = (u_2, u_3)) \\
&\quad\quad + (\text{The total effect of } U_2 = u_2 \text{ on } Y = y \text{ at } (U_1, U_3) = (u_1, u_3)) \\
&\quad\quad + (\text{The total effect of } U_3 = u_3 \text{ on } Y = y \text{ at } (U_1, U_2) = (u_1, u_2))
\end{aligned} \tag{7.18}$$

Remark 7.2 The effects defined in this section are interpreted in information, and the exponentials of them are viewed as the multiplicative effects in odds ratios. ■

In order to summarize and standardize the effects based on log odds ratios, the entropy coefficient of determination (ECD) [6] is used. In logit model (7.9), the standardized summary total effect of (U_1, U_2, U_3) on Y is given by

$$e_T((U_1, U_2, U_3) \to Y) = \frac{\sum_{k=1}^{3} \beta_k \mathrm{Cov}(Y, U_k)}{\sum_{k=1}^{3} \beta_k \mathrm{Cov}(Y, U_k) + 1}. \tag{7.19}$$

Remark 7.3 By taking the expectation of (7.11) over all (u_1, u_2, u_3) and y, we have $\sum_{k=1}^{3} \beta_k \mathrm{Cov}(Y, U_k)$ and (7.19) is ECD of (U_1, U_2, U_3) and Y. ■

Summarizing and standardizing the effects from (7.12) to (7.17) as in (7.19), we also have

$$e_T((U_2, U_3) \to Y) = \frac{\sum_{k=2}^{3} \beta_k \mathrm{Cov}(Y, U_k | U_1)}{\sum_{k=1}^{3} \beta_k \mathrm{Cov}(Y, U_k) + 1},$$

$$e_T(U_1 \to Y) = \frac{\sum_{k=1}^{3} \beta_k \mathrm{Cov}(Y, U_k) - \sum_{k=2}^{3} \beta_k \mathrm{Cov}(Y, U_k | U_1)}{\sum_{k=1}^{3} \beta_k \mathrm{Cov}(Y, U_k) + 1},$$

$$e_D(U_1 \to Y) = \frac{\beta_1 \mathrm{Cov}(Y, U_1 | U_2, U_3)}{\sum_{k=1}^{3} \beta_k \mathrm{Cov}(Y, U_k) + 1},$$

$$e_T(U_3 \to Y) = e_D(U_3 \to Y) = \frac{\beta_3 \mathrm{Cov}(Y, U_3 | U_1, U_2)}{\sum_{k=1}^{3} \beta_k \mathrm{Cov}(Y, U_k) + 1},$$

$$e_T((U_2, U_3) \to Y) = \frac{\sum_{k=2}^{3} \beta_k \mathrm{Cov}(Y, U_k | U_1)}{\sum_{k=1}^{3} \beta_k \mathrm{Cov}(Y, U_k) + 1},$$

$$e_T(U_2 \to Y) = e_T((U_2, U_3) \to Y) - e_T(U_3 \to Y),$$

$$e_D(U_2 \to Y) = \frac{\beta_2 \mathrm{Cov}(Y, U_2|U_1, U_3)}{\sum_{k=1}^{3} \beta_k \mathrm{Cov}(Y, U_k) + 1},$$

where notations $e_D(*)$ and $e_T(*)$ imply the standardized summary direct and total effects of the related variables, respectively. In the above path analysis, from (7.18), we also have

$$e_T((U_1, U_2, U_3) \to Y) = \sum_{k=1}^{3} e_T(U_k \to Y).$$

Next, path analysis for a path system in Fig. 7.4 is carried out. The joint probability of the three variables is decomposed as follows:

$$f(u_1, u_2, y) = f_{12}(u_1, u_2) f(y|u_1, u_2),$$

where $f_{12}(u_1, u_2)$ is the joint probability function of U_1 and U_2 and $f(y|u_1, u_2)$ the conditional probability function of Y for given $(U_1, U_2) = (u_1, u_2)$. For the following logit model:

$$f(y|u_1, u_2) = \frac{\exp(y\alpha + y\beta_1 u_1 + y\beta_2 u_2)}{1 + \exp(\alpha + y\beta_1 u_1 + y\beta_2 u_2)},$$

the total effect of $(U_1, U_2) = (u_1, u_2)$ on $Y = y$ is given by

$$(y - \nu)\beta_1(u_1 - \mu_1) + (y - \nu)\beta_2(u_2 - \mu_2).$$

The direct effect of $U_1 = u_1$ on $Y = y$ at $U_2 = u_2$ and that of $U_2 = u_2$ on $Y = y$ at $U_1 = u_1$ are given, respectively, as follows:

$$(y - \nu(u_2))\beta_1(u_1 - \mu_1(u_2))$$

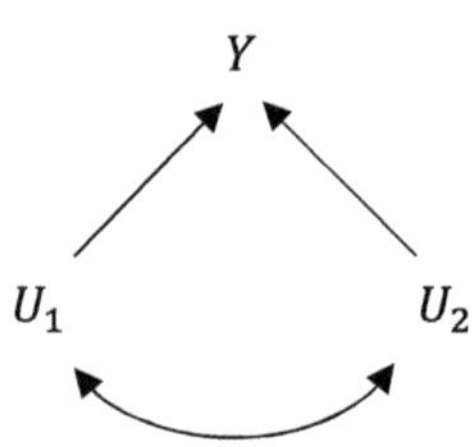

Fig. 7.4 A path diagram of Y and its parent variables $U_i, i = 1, 2$

and

$$(y - \nu(u_1))\beta_2(u_2 - \mu_2(u_1)),$$

where $\nu(u_k), k = 1,2, \mu_1(u_2)$ and $\nu(u_1)$ are the conditional expectations of Y, U_1, and U_2, respectively, as in the above discussion. The total effect of $U_1 = u_1$ on $Y = y$ at $U_2 = u_2$ is calculated by

$$\begin{aligned}
&(\text{the total effect of } (U_1, U_2) = (u_1, u_2) \text{ on } Y = y) \\
&\quad - (\text{the direct effect of } U_2 = u_2 \text{ on } Y = y \text{ at } U_1 = u_1) \\
&= (y - \nu)\beta_1(u_1 - \mu_1) + (y - \nu)\beta_2(u_2 - \mu_2) - (y - \nu(u_1))\beta_2(u_2 - \mu_2(u_1)).
\end{aligned}$$

Similarly, the total effect of $U_2 = u_2$ on $Y = y$ at $U_1 = u_1$ can also be calculated. Summarizing and standardizing the above effects, we have

$$e_T((U_1, U_2) \to Y) = \frac{\beta_1 \mathrm{Cov}(Y, U_1) + \beta_2 \mathrm{Cov}(Y, U_2)}{1 + \beta_1 \mathrm{Cov}(Y, U_1) + \beta_2 \mathrm{Cov}(Y, U_2)},$$

$$e_D(U_1 \to Y) = \frac{\beta_1 \mathrm{Cov}(Y, U_1|U_2)}{1 + \beta_1 \mathrm{Cov}(Y, U_1) + \beta_2 \mathrm{Cov}(Y, U_2)},$$

$$e_D(U_2 \to Y) = \frac{\beta_2 \mathrm{Cov}(Y, U_2|U_1)}{1 + \beta_1 \mathrm{Cov}(Y, U_1) + \beta_2 \mathrm{Cov}(Y, U_2)},$$

$$e_T(U_1 \to Y) = e_T((U_1, U_2) \to Y) - e_D(U_2 \to Y),$$

$$e_T(U_2 \to Y) = e_T((U_1, U_2) \to Y) - e_D(U_1 \to Y).$$

The above method can be applied to the causal system in Fig. 7.2b.

7.4 Path Analysis in Multiple-Indicator, Multiple-Cause Models

7.4.1 The Multiple-Indicator, Multiple-Cause Model in Fig. 7.2a

The above method of path analysis is applied to a multiple-cause, multiple-indicator model in Fig. 7.2a. According to (7.6), the effects of S_1 on $X_{1i}, i = 1,2, \ldots, I_1$ and those of S_1 and S_2 on $X_{2i}, i = 1,2, \ldots, I_2$ are calculated. Let ν_k and μ_{ki} be the expectations of X_{ki} and S_k, respectively. Since the variables concerned are binary, we have

$$\mu_k = P(S_k = 1), k = 1{,}2;$$

$$\nu_{ki} = P(X_{ki} = 1), i = 1{,}2, \ldots, I_k, k = 1{,}2.$$

(i) The effects of S_1 on X_{1i}

For $S_1 \to X_{1i}$, applying the discussion in Sect. 7.3 to (7.7), by using a method similar to (7.11) we have the total (direct) effects of $S_1 = s_1$ on $X_{1i} = x_{1i}$ as

$$(x_{1i} - \nu_{1i})\beta_{1i}(s_1 - \mu_1), i = 1{,}2, \ldots, I_1. \quad (7.20)$$

Summarizing and standardizing the above effects, we have

$$e_T(S_1 \to X_{1i}) = \frac{\beta_{1i}\mathrm{Cov}(X_{1i}, S_1)}{\beta_{1i}\mathrm{Cov}(X_{1i}, S_1) + 1}, i = 1{,}2, \ldots, I_1. \quad (7.21)$$

(ii) The effects of S_1 on S_2

In a way similar to (7.21), in (7.8) we have

$$''\text{the total (direct) effect of } S_1 = s_1 \text{ on } S_2 = s_2'' = (s_2 - \mu_2)\delta_1(s_1 - \mu_1), \quad (7.22)$$

$$e_T(S_1 \to S_2) = \frac{\delta_1\mathrm{Cov}(S_1, S_2)}{\delta_1\mathrm{Cov}(S_1, S_2) + 1}. \quad (7.23)$$

(iii) The effects of S_1 and S_2 on X_{2i}

From logit model (7.7), the total effects of $(S_1, S_2) = (s_1, s_2)$ on $X_{2i} = x_{2i}$ are given by

$$(x_{2i} - \nu_{2i})\beta_{2i}(s_2 - \mu_2), i = 1{,}2, \ldots, I_2. \quad (7.24)$$

According to (7.12), the total (direct) effects of $S_2 = s_2$ on $X_{2i} = x_{2i}$ at $S_1 = s_1$ are calculated as follows:

$$(x_{2i} - \nu_{2i}(s_1))\beta_{2i}(s_2 - \mu_2(s_1)), i = 1{,}2, \ldots, I_2, \quad (7.25)$$

where $\nu_2(s_1)$ and $\mu_{2i}(s_1)$ are the conditional expectation of X_{2i} and S_2 given $S_1 = s_1$, respectively. The variables are binary, so it follows that

$$\mu_2(s_1) = P(S_2 = 1|S_1 = s_1), \nu_{2i}(s_1) = P(X_{2i} = 1|S_1 = s_1).$$

Since the direct effects of $S_1 = s_1$ on $X_{2i} = x_{2i}$ are zero, by subtracting (7.25) from (7.24), the total (indirect) effects of $S_1 = s_1$ on $X_{2i} = x_{2i}$ through $S_2 = s_2$ are obtained as follows:

$$(x_{2i} - \nu_{2i})\beta_{2i}(s_2 - \mu_2) - (x_{2i} - \nu_{2i}(s_1))\beta_{2i}(s_2 - \mu_2(s_1)), i = 1,2, \ldots, I_2.$$

Summarizing and standardizing the above effects, we have

$$e_T((S_1, S_2) \to X_{2i}) = \frac{\beta_{2i}\mathrm{Cov}(X_{2i}, S_2)}{1 + \beta_{2i}\mathrm{Cov}(X_{2i}, S_2)}, i = 1,2, \ldots, I_2, \quad (7.26)$$

$$e_T(S_2 \to X_{2i}) = e_D(S_2 \to X_{2i}) = \frac{\beta_{2i}\mathrm{Cov}(X_{2i}, S_2|S_1)}{1 + \beta_{2i}\mathrm{Cov}(X_{2i}, S_2)}, i = 1,2, \ldots,, \quad (7.27)$$

$$\begin{aligned} e_T(S_1 \to X_{2i}) &= e_I(S_1 \to X_{2i}) = e_T((S_1, S_2) \to X_{2i}) - e_T(S_2 \to X_{2i}) \\ &= \frac{\beta_{2i}\mathrm{Cov}(X_{2i}, S_2) - \beta_{2i}\mathrm{Cov}(X_{2i}, S_2|S_1)}{1 + \beta_{2i}\mathrm{Cov}(X_{2i}, S_2)}, i = 1,2, \ldots, I_2. \end{aligned}$$

More complicated models as shown in Fig. 7.5 can also be considered. The partial path system of S_1, S_2, (X_{1i}), and (X_{2i}) can be analyzed as above, so it is sufficient to discuss the partial path system of S_1, S_2, S_3, and (X_{3i}) as shown in Fig. 7.6. The diagram is a special case of Fig. 7.3. Hence, the discussion on the path diagram in Fig. 7.3 can be directly employed.

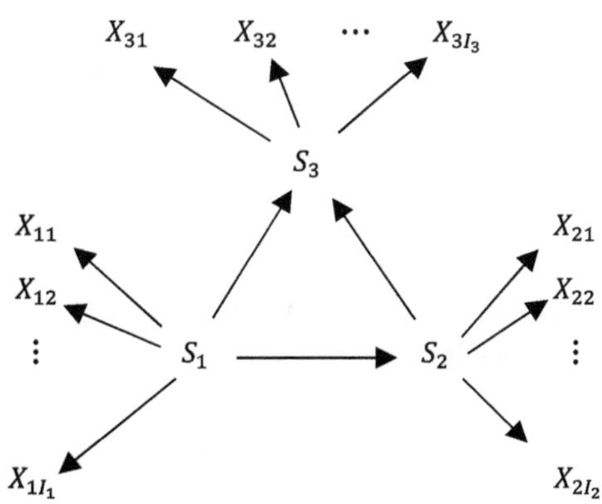

Fig. 7.5 Path diagram of manifest variables $X_{ki}, i = 1,2, \ldots, I_k$ and latent variables $S_k, i = 1,2,3$

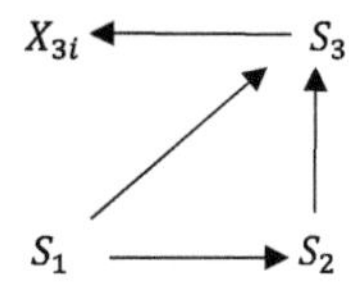

Fig. 7.6 Partial path diagram of manifest variables X_{3i} and latent variables $S_k, i = 1,2,3$

7.4.2 *The Multiple-Indicator, Multiple-Cause Model in Fig. 7.2b*

For the path diagram in Fig. 7.2b, the effects of variables are calculated as follows:

(i) The effects of S_1 and S_2 on X_{1i}

From logit model (7.7), the total effects of $(S_1, S_2) = (s_1, s_2)$ on X_{1i} are given by

$$(x_{1i} - \nu_{1i})\beta_{1i}(s_1 - \mu_1), i = 1,2, \ldots, I_1. \tag{7.28}$$

Since the direct effects of $S_2 = s_2$ on $X_{1i} = x_{1i}$ are zero, the above effects are also the total effects of $S_1 = s_1$. With a method similar to (iii) in Subsection 7.4.1, the direct effects of $S_1 = s_1$ on $X_{1i} = x_{1i}$ at $S_2 = s_2$ are given as follows:

$$(x_{1i} - \nu_{1i}(s_2))\beta_{1i}(s_1 - \mu_1(s_2)), i = 1,2, \ldots, I_1, \tag{7.29}$$

where $\nu_1(s_2)$ and $\mu_{1i}(s_2)$ are the conditional expectation of X_{1i} and S_1 given $S_2 = s_2$, that is,

$$\mu_1(s_2) = P(S_1 = 1|S_2 = s_2), \nu_{1i}(s_2) = P(X_{1i} = 1|S_2 = s_2). \tag{7.30}$$

By subtracting (7.29) from (7.28), the indirect effects of $S_1 = s_1$ on $X_{1i} = x_{1i}$ through the association with $S_2 = s_2$ are obtained as follows:

$$(x_{1i} - \nu_{1i})\beta_{1i}(s_1 - \mu_1) - (x_{1i} - \nu_{1i}(s_2))\beta_{1i}(s_1 - \mu_1(s_2)), i = 1,2, \ldots, I_1.$$

The above effects are also the indirect effects of $S_2 = s_2$ on $X_{1i} = x_{1i}$ as well. Summarizing and standardizing the above effects, we have

$$e_T((S_1, S_2) \to X_{1i}) = e_T(S_1 \to X_{1i}) = \frac{\beta_{1i}\mathrm{Cov}(X_{1i}, S_1)}{1 + \beta_{1i}\mathrm{Cov}(X_{1i}, S_1)}, i = 1,2, \ldots, I_1;$$

$$e_D(S_1 \to X_{1i}) = \frac{\beta_{1i}\mathrm{Cov}(X_{1i}, S_1|S_2)}{1 + \beta_{1i}\mathrm{Cov}(X_{1i}, S_1)}, i = 1,2, \ldots, I_1; \tag{7.31}$$

$$\begin{aligned} e_I(S_1 \to X_{1i})(= e_I(S_2 \to X_{1i})) &= e_T(S_1 \to X_{1i}) - e_D(S_1 \to X_{1i}) \\ &= \frac{\beta_{1i}\mathrm{Cov}(X_{1i}, S_1) - \beta_{1i}\mathrm{Cov}(X_{1i}, S_1|S_2)}{1 + \beta_{1i}\mathrm{Cov}(X_{1i}, S_1)}, i = 1,2, \ldots, I_1. \end{aligned}$$

$$e_D(S_2 \to X_{1i}) = 0, i = 1,2, \ldots, I_1.$$

(ii) The effects of S_1 and S_2 on X_{2i}

By a method similar to the calculation of the effects of S_1 and S_2 on X_{1i}, substituting S_1, S_2, and X_{1i} in (i) for S_2, S_1, and X_{2i}, respectively, we can obtain the effects of S_1 and S_2 on X_{2i}.

7.5 Numerical Illustration I

7.5.1 *Model I (Fig. 7.2a)*

Table 7.1 shows artificial parameters of a multiple-indicator, multiple-cause model with two binary latent variables and three indicator manifest variables for each latent variable, for demonstrating the path analysis in Fig. 7.7. By using the parameters, we can get the regression coefficients β and δ in (7.21) and (7.23), as shown in Tables 7.2, respectively. Table 7.3 illustrates the means of latent variables S_t and manifest ones X_{ti}, and the conditional means of S_2 and X_{2i} given $S_1 = 0$ or 1 are given in Table 7.4. By using the parameters, the effects of variables are calculated. First, the effects of latent variable S_1 on manifest variables $X_{1i}, i = 1,2,3$ are obtained in Table 7.5. According to the path diagram in Fig. 7.7, the effects are direct ones, and also total ones, that is,

$$\text{the total effects of } \mathrm{S}_1 = 1 \text{ on } \mathrm{X}_{1\mathrm{i}} = \text{the direct effects of } \mathrm{S}_1 = 1 \text{ on } \mathrm{X}_{1\mathrm{i}}.$$

From the table, for example,

Table 7.1 Parameters of a multiple-indicator, multiple-cause model in Fig. 7.7

Latent class	Proportion	Latent positive item response probability					
		X_{11}	X_{12}	X_{13}	X_{21}	X_{22}	X_{23}
(1,1)	0.32	0.9	0.7	0.6	0.8	0.8	0.7
(0,1)	0.18	0.2	0.1	0.3	0.8	0.8	0.7
(1,0)	0.08	0.9	0.7	0.6	0.1	0.3	0.1
(0,0)	0.42	0.2	0.1	0.3	0.1	0.3	0.1

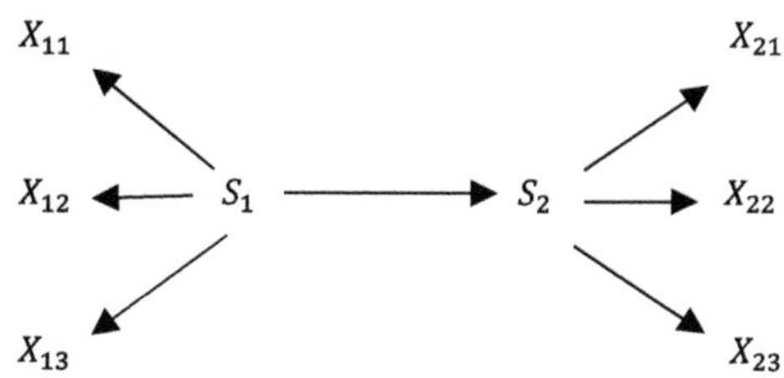

Fig. 7.7 Path diagram of manifest and latent variables for Numerical Illustration I

Table 7.2 The estimated regression coefficients in (7.28), (7.30), and (7.32)

δ	β_{11}	β_{12}	β_{13}	β_{21}	β_{22}	β_{23}
2.234	3.584	3.045	1.253	3.584	2.234	3.045

Table 7.3 The means of variables latent and manifest variables, S_t and X_{ti}

μ_1	μ_2	ν_{11}	ν_{12}	ν_{13}	ν_{21}	ν_{22}	ν_{23}
0.4	0.5	0.48	0.34	0.42	0.45	0.55	0.4

Table 7.4 The conditional means of S_2 and X_{2i} given $S_1 = 0\text{or}1$

s_1	$\mu_2(s_1)$	$\nu_{21}(s_1)$	$\nu_{22}(s_1)$	$\nu_{23}(s_1)$
1	0.8	0.66	0.7	0.58
0	0.3	0.31	0.45	0.28

Table 7.5 The total (direct) effects of Latent variable S_1 on Manifest variables $X_{1i}, i = 1,2,3$

S_1	X_{11}	X_{12}	X_{13}	$S_1 \to X_{11}$	$S_1 \to X_{12}$	$S_1 \to X_{13}$
1	1	1	1	1.118	1.206	0.436
0	1	1	1	−0.745	−0.804	−0.291
1	0	0	0	−1.032	−0.621	−0.316
0	0	0	0	0.688	0.414	0.210
Mean effect $\beta_{1i}\text{Cov}(X_{1i}, S_1)$				0.602	0.438	0.090
$e_T(S_1 \to X_{1i}) = e_D(S_1 \to X_{1i})$				0.376	0.305	0.083

$$
\begin{aligned}
&\text{the total effects of } S_1 = 1 \text{ on } X_{11} = 1 = 1.118,\\
&\text{the total effects of } S_1 = 1 \text{ on } X_{12} = 1 = 1.206,\\
&\text{the total effects of } S_1 = 1 \text{ on } X_{13} = 1 = 0.436.
\end{aligned}
$$

The above effects are the changes of information in X_{1i} for latent variable $S_1 = 1$ as explained in (7.10), and the following exponentials of the quantities can be interpreted as the odds ratios:

$$
\exp(1.118) = 3.058, \exp(1.206) = 3.340, \exp(0.436) = 1.547.
$$

For baselines ν_1 and $\mu_{1i}, i = 1,2,3$ in Table 7.3, the odds ratios with respect to the variables $(S_1, X_{1i}), i = 1,2,3$ are 3.058, 3.340, and 1.547, respectively. The standardized effects (7.21) are shown in the seventh row of Table 7.5. For example, $e_T(S_1 \to X_{11}) = 0.376$ implies that 37.6% of the variation of manifest variable X_{11} in entropy is explained by latent variable S_1. By using (7.22) and (7.23), the effects of latent variable S_1 on S_2 are calculated in Table 7.6, and the explanation of the table can be given as in Table 7.5. In Table 7.7, the total effects of latent variables (S_1, S_2)

Table 7.6 The total (direct) effects of latent variable S_1 on S_2

S_1	S_2	$S_1 \to S_2$
1	1	−0.137
0	1	0.206
1	0	0.303
0	0	−0.104
Mean effect $\delta_1 \mathrm{Cov}(S_1, S_2)$		0.268
$e_T(S_1 \to S_2) = e_D(S_1 \to S_2)$		0.211

Table 7.7 The total effects of latent variables (S_1, S_2) on manifest variables $X_{2i}, i = 1,2, 3$

S_1	S_2	X_{21}	X_{22}	X_{23}	$(S_1, S_2) \to X_{21}$	$(S_1, S_2) \to X_{22}$	$(S_1, S_2) \to X_{23}$
1	1	1	1	1	0.985	0.503	0.913
0	1	1	1	1	0.985	0.503	0.913
1	0	1	1	1	−0.985	−0.503	−0.913
0	0	1	1	1	−0.985	−0.503	−0.913
1	1	0	0	0	−0.806	−0.614	−0.609
0	1	0	0	0	−0.806	−0.614	−0.609
1	0	0	0	0	0.806	0.614	0.609
0	0	0	0	0	0.806	0.614	0.609
Mean effect $\beta_{2i} \mathrm{Cov}(X_{2i}, S_2)$					0.627	0.279	0.457
$e_T((S_1, S_2) \to X_{2i})$					0.385	0.218	0.314

on manifest variables $X_{2i}, i = 1,2, 3$ are calculated. According to the path diagram in Fig. 7.7, latent variable S_1 and manifest variables X_{2i} are conditionally independent, given S_2, so the total effects of $(S_1, S_2) = (1, s), s = 0,1$ on $X_{2i}, i = 1,2, 3$ are equal to those of $(S_1, S_2) = (0, s)$ on X_{2i}, that is, the effects depend only on S_2, as shown in Table 7.7. The effects are calculated according to (7.24), for example,

$$\text{'the total effects of } (S_1, S_2) = (s, 1) \text{ on } X_{21} = 1' = 0.985, \ s = 0, 1;$$
$$\text{'the total effects of } (S_1, S_2) = (s, 0) \text{ on } X_{21} = 1' = -0.985, \ s = 0, 1;$$
$$\text{'the total effects of } (S_1, S_2) = (s, 1) \text{ on } X_{21} = 0' = 0.806, \ s = 0, 1;$$
$$\text{'the total effects of } (S_1, S_2) = (s, 0) \text{ on } X_{21} = 0' = 0.806, s = 0, 1.$$

The exponentials of the above effects can be interpreted as odds ratios as in Table 7.5. The standardized effects are obtained through (7.26).

Remark 7.4 In Table 7.7, the absolute values of effects of $(S_1, S_2) = (i, j), i = 0,1$ on X_{2k} are the same for $j = 0,1; k = 1,2, 3$, because of (7.24) and the mean of $S_2(= \mu_2) = \frac{1}{2}$ (Table 7.3). ■

Table 7.8 shows the total (direct) effects of latent variable S_2 on manifest variables $X_{2i}, i = 1,2, 3$. The effects are calculated with (7.24) and Table 7.4. The standardized

Table 7.8 The total (direct) effects of latent variable S_2 on manifest variables $X_{2i}, i = 1,2,3$

S_1	S_2	X_{21}	X_{22}	X_{23}	$S_2 \to X_{21}$	$S_2 \to X_{22}$	$S_2 \to X_{23}$
1	1	1	1	1	0.244	0.134	0.256
1	1	0	0	0	−0.473	−0.313	−0.353
1	0	1	1	1	−0.975	−0.536	−1.023
1	0	0	0	0	1.892	1.251	1.413
0	1	1	1	1	1.731	0.860	1.534
0	1	0	0	0	−0.778	−0.704	−0.597
0	0	1	1	1	−0.742	−0.369	−0.658
0	0	0	0	0	0.333	0.302	0.256
Mean effect $\beta_{2i}\mathrm{Cov}(X_{2i}, S_2\vert S_1)$					0.477	0.212	0.347
$e_T(S_2 \to X_{2i})$					0.293	0.166	0.238

Table 7.9 The indirect effects of latent variable S_1 on manifest variables $X_{2i}, i = 1,2,3$

S_1	S_2	X_{21}	X_{22}	X_{23}	$S_2 \to X_{21}$	$S_2 \to X_{22}$	$S_2 \to X_{23}$
1	1	1	1	1	0.741	0.369	0.657
1	1	0	0	0	−0.333	−0.301	−0.256
1	0	1	1	1	−0.011	0.033	0.110
1	0	0	0	0	−1.086	−0.637	−0.804
0	1	1	1	1	−0.746	−0.357	−0.621
0	1	0	0	0	−0.028	0.090	−0.012
0	0	1	1	1	−0.243	−0.134	−0.256
0	0	0	0	0	0.473	0.312	0.353
Mean effect $\beta_{2i}\mathrm{Cov}(X_{2i}, S_2) - \beta_{2i}\mathrm{Cov}(X_{2i}, S_2\vert S_1)$					0.151	0.067	0.110
$e_T(S_1 \to X_{2i}) = e_I(S_1 \to X_{2i})$					0.093	0.052	0.075

effects are given by (7.27). By subtracting Table 7.8 from Table 7.7, we have the indirect effects of latent variable S_1 on manifest variables $X_{2i}, i = 1,2,3$, as shown in Table 7.9.

7.5.2 *Model II (Fig. 7.2b)*

In order to compare the results of path analysis in Figs. 7.7 and 7.8, the same parameters in Table 7.1 are used. In this case, the total effects of latent variable S_1 on manifest variables $X_{1i}, i = 1,2,3$ are the same as in Table 7.5; however, the direct effects of S_1 are calculated according to (7.29), (7.30), and (7.31) and we have Table 7.10. Since according to Fig. 7.8, the total effects of (S_1, S_2) on X_{1i} are the same

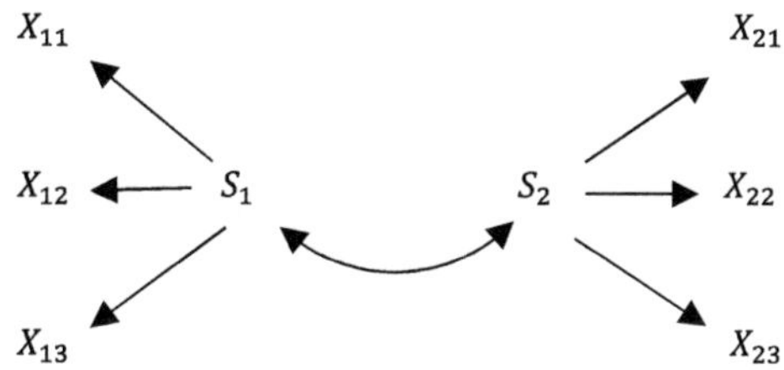

Fig. 7.8 Path diagram of manifest and latent variables for Numerical Illustration I

Table 7.10 The direct effects of latent variable S_1 on manifest variables $X_{1i}, i = 1,2,3$

S_2	S_1	X_{11}	X_{12}	X_{13}	$S_1 \to X_{11}$	$S_1 \to X_{12}$	$S_1 \to X_{13}$
1	1	1	1	1	0.245	0.228	0.191
1	0	1	1	1	−0.436	−0.974	−0.292
1	1	0	0	0	−1.045	−0.350	−0.304
1	0	0	0	0	1.858	1.492	0.466
0	1	1	1	1	2.228	1.885	0.744
0	0	1	1	1	−0.424	−0.662	−0.145
0	1	0	0	0	−0.783	−0.368	−0.304
0	0	0	0	0	0.149	0.129	0.059
Mean effect $\beta_{1i}\mathrm{Cov}(X_{1i}, S_1 \vert S_2)$					0.458	0.340	0.066
$e_D(S_1 \to X_{1i})$					0.286	0.250	0.060

as those of S_1 on X_{1i}, by subtracting Table 7.10 from Table 7.5, we have the indirect effects of S_1 on X_{1i}, which are shown in Table 7.11. The effects are also those of S_2 on X_{1i}. Similarly, the total effects of latent variable S_2 on manifest variables $X_{2i}, i = 1,2,3$ are the same as those of (S_1, S_2) shown in Table 7.7, and are given in

Table 7.11 The indirect effects of latent variable S_1 (S_2) on manifest variables $X_{1i}, i = 1,2,3$

S_2	S_1	X_{11}	X_{12}	X_{13}	$S_1 \to X_{11}$	$S_1 \to X_{12}$	$S_1 \to X_{13}$
1	1	1	1	1	0.873	0.977	0.245
1	0	1	1	1	−0.310	0.170	0.001
1	1	0	0	0	2.163	1.556	0.740
1	0	0	0	0	−2.603	−2.296	−0.757
0	1	1	1	1	−3.260	−2.506	−1.060
0	0	1	1	1	1.112	1.076	0.355
0	1	0	0	0	−0.249	−0.253	−0.012
0	0	0	0	0	0.539	0.285	0.151
Mean effect $\beta_{1i}\mathrm{Cov}(X_{1i}, S_1) - \beta_{1i}\mathrm{Cov}(X_{1i}, S_1 \vert S_2)$					0.144	0.079	0.024
$e_I(S_1 \to X_{1i}) = e_I(S_2 \to X_{1i})$					0.090	0.055	0.022

Table 7.12 The total effects of latent variables S_2 on manifest variables $X_{2i}, i = 1,2,3$

S_2	X_{21}	X_{22}	X_{23}	$S_2 \to X_{21}$	$S_2 \to X_{22}$	$S_2 \to X_{13}$
1	1	1	1	0.985	0.503	0.913
0	1	1	1	−0.985	−0.503	−0.913
1	0	0	0	−0.806	−0.614	−0.609
0	0	0	0	0.806	0.614	0.609
Mean effect $\beta_{2i}\mathrm{Cov}(X_{2i}, S_2)$				0.627	0.279	0.457
$e_T(S_2 \to X_{2i})$				0.385	0.218	0.314

Table 7.12. The direct effects of S_2 on $X_{2i}, i = 1,2,3$ are the same as those in Table 7.8. The indirect effects of S_1 on $X_{2i}, i = 1,2,3$ are the same as those calculated in Table 7.9, and the effects are also the indirect effects of S_2 on $X_{2i}, i = 1,2,3$, based on Fig. 7.8.

The above method is applied to McHugh data, which were on test on creative ability in machine design (Chap. 2). Assuming latent skills S_i for solving subtests $X_i, i = 1,2,3,4$, a confirmatory latent class model for explaining a learning structure is used, and the results of the analysis are given in Table 4.2. From the results, two latent skills $S_1(= S_2)$ and $S_3(= S_4)$ for solving the test can be assumed. In Chap. 4, assuming learning processes in a population, a path analysis has been performed. In this section, the model is viewed as a multiple-indicator, multiple-cause model, and the present method of path analysis is applied. The path diagram of the manifest and latent variables is shown in Fig. 7.9. By using the present approach, we have the mean effects (Table 7.13). From the table, the mean total effects of (S_1, S_2) on X_1 and X_2 are equal to those of S_1, and the mean total effects of (S_1, S_2) on X_3 and X_4 are equal to those of S_2. The indirect effects of S_1 on $X_i, i = 1,2,3,4$ are equal to those of S_2. Thus, in the path diagram in Fig. 7.9, the indirect effects are induced by the association between the latent variables S_1 on S_2. Using the mean effects in Table 7.13, the standardized effects are calculated according to the present method (Table 7.14). In order to interpret the effects based on entropy as ECD, Table 7.14 illustrates the standardized effects of the mean effects shown in Table 7.13. As shown in the table, the indirect effects are relatively small, for example, in the standardized effects of S_1 on X_1 and X_2, the indirect effects are moderate, and about 1/4 times the direct effects.

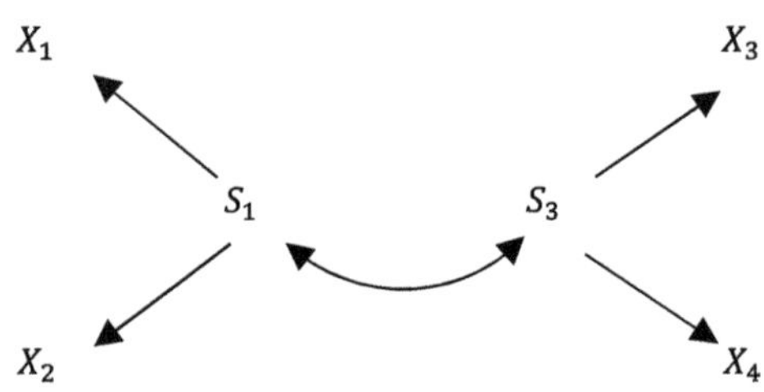

Fig. 7.9 Path diagram of manifest and latent variables for McHugh data

Table 7.13 Mean effects of latent variables S_1 and S_3 on manifest variables X_i, $i = 1,2,3,4$

Mean effect	Manifest variable			
	X_{11}	X_{21}	X_{12}	X_{22}
Total effect of S_1 and S_2	0.519	1.204	1.278	0.454
Total effect of S_1	0.519	1.204	0.277	0.099
Direct effect of S_1	0.406	0.943	0	0
Indirect effect of S_1	0.113	0.261	0.277	0.009
Total effect of S_2	0.113	0.261	1.278	0.454
Direct effect of S_2	0	0	1.001	0.356
Indirect effect of S_2	0.113	0.261	0.277	0.009

Table 7.14 Standardized effects of latent variables S_1 and S_3 on manifest variables X_i, $i = 1,2,3,4$

Standardized effect	Manifest variable			
	X_{11}	X_{21}	X_{12}	X_{22}
Total effect of S_1 and S_2	0.342	0.546	0.561	0.312
Total effect of S_1	0.342	0.546	0.122	0.068
Direct effect of S_1	0.268	0.428	0	0
Indirect effect of S_1	0.074	0.118	0.122	0.068
Total effect of S_2	0.074	0.118	0.561	0.312
Direct effect of S_2	0	0	0.439	0.245
Indirect effect of S_2	0.074	0.118	0.122	0.068

7.6 Path Analysis of the Latent Markov Chain Model

The present path analysis is applied to the latent Markov chain model treated in Chap. 5. As in Sect. 5.2 in Chap. 5, let X_t be manifest variables that take values on sample space $\Omega_{manifest} = \{1,2,\ldots,J\}$ at time points $t = 1,2,\ldots$, and let S_t be the corresponding latent variables on sample space $\Omega_{latent} = \{1,2,\ldots,A\}$, which are assumed a first-order time-homogeneous Markov chain. Let $m_{ab}, a, b = 1,2,\ldots,A$ be the transition probabilities; let $q_a, a = 1,2,\ldots,A$ be the probabilities of $S_1 = a$, that is, the initial state distribution; and let p_{aj} be the probabilities of $X_t = j$, given $S_t = a$, that is, $p_{aj} = P(X_t = j|S_t = a)$ and the probabilities are independent of time t. In order to make a general discussion, the following dummy variables for manifest and latent categories are introduced. Let

$$X_{tj} = \begin{cases} 1 \text{ for } X_t = j, \\ 0 \text{ otherwise,} \end{cases} \quad \text{and} \quad S_{ta} = \begin{cases} 1 \text{ for } S_t = a, \\ 0 \text{ otherwise.} \end{cases}$$

Then, manifest and latent variables X_t and S_t are identified to the following dummy variable vectors, respectively:

$$\boldsymbol{X}_t = (X_{t1}, X_{t2}, \ldots, X_{tJ})^T \quad \text{and} \quad \boldsymbol{S}_t = (S_{t1}, S_{t2}, \ldots, S_{tA})^T.$$

For convenience of the discussion, based on the above identification, transition probabilities m_{ab} and response probabilities p_{aj} are expressed as $m_{s_{t-1}s_t}$ and $p_{s_t x_t}$, respectively, that is, if dummy state vectors $\boldsymbol{s}_{t-1}$ and $\boldsymbol{s}_t$ have elements $s_{t-1,a} = 1$ and $s_{tb} = 1$, respectively, it implies that $m_{s_{t-1}s_t} = m_{ab}$; and if dummy state vector $\boldsymbol{s}_t$ and dummy response vector $\boldsymbol{x}_t$ have elements $s_{ta} = 1$ and $x_{tj} = 1$, it means that $p_{s_t x_t} = p_{aj}$. Let the transition matrix and the response matrix be denoted by

$$\boldsymbol{M} = (m_{ab}) \text{ and } \boldsymbol{P} = \left(p_{aj}\right),$$

respectively. Then, the probabilities are re-expressed as follows:

$$p_{s_t x_t} = \frac{\exp\left(\boldsymbol{\alpha x}_t + \boldsymbol{s}_t^T \mathbf{B} \boldsymbol{x}_t\right)}{\sum_{s_t} \exp\left(\boldsymbol{\alpha x}_t + \boldsymbol{s}_t^T \mathbf{B} \boldsymbol{x}_t\right)} \quad \text{and} \quad m_{s_{t-1}s_t} = \frac{\exp\left(\boldsymbol{\gamma s}_t + \boldsymbol{s}_{t-1}^T \boldsymbol{\Delta s}_t\right)}{\sum_{s_t} \exp\left(\boldsymbol{\gamma s}_t + \boldsymbol{s}_{t-1}^T \boldsymbol{\Delta s}_t\right)},$$

where

$$\boldsymbol{\alpha} = \begin{pmatrix} \alpha_1 \\ \alpha_2 \\ \vdots \\ \alpha_J \end{pmatrix}, \mathbf{B} = \begin{pmatrix} \beta_{11} & \beta_{12} & \cdots & \beta_{1J} \\ \beta_{21} & \beta_{22} & \cdots & \beta_{2J} \\ \vdots & \vdots & \ddots & \vdots \\ \beta_{A1} & \beta_{A1} & \cdots & \beta_{AJ} \end{pmatrix}, \tag{7.32}$$

$$\boldsymbol{\gamma} = \begin{pmatrix} \gamma_1 \\ \gamma_2 \\ \vdots \\ \gamma_A \end{pmatrix}, \boldsymbol{\Delta} = \begin{pmatrix} \delta_{11} & \delta_{12} & \cdots & \delta_{1A} \\ \delta_{21} & \delta_{22} & \cdots & \delta_{2A} \\ \vdots & \vdots & \ddots & \vdots \\ \delta_{A1} & \delta_{A1} & \cdots & \delta_{AA} \end{pmatrix}. \tag{7.33}$$

Figure 7.10 shows the path diagram of the latent Markov chain model treated above. According to the model, the following sequence is a Markov chain:

$$S_1 \to S_2 \to S_3 \to \cdots \to S_t \to X_t. \tag{7.34}$$

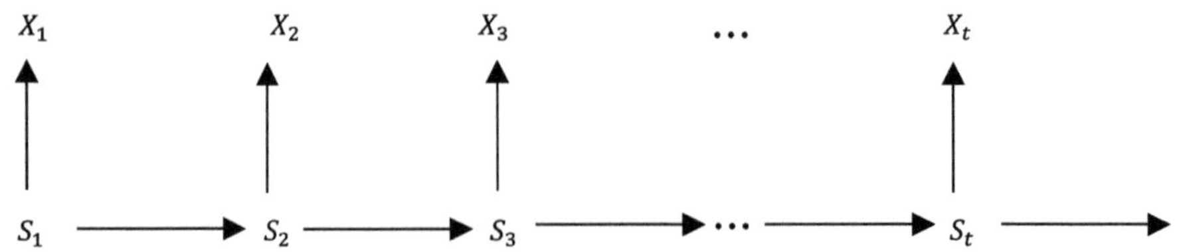

Fig. 7.10 The latent Markov chain model

First, the effects of latent variables $\boldsymbol{S}_u, u = 1,2, \ldots, t$ on manifest variable $\boldsymbol{X}_t$ are discussed. For simplicity of the discussion, setting $t = 3$, the effects of the latent variables are calculated. According to path analysis [9], the total effect of $(\boldsymbol{S}_1, \boldsymbol{S}_2, \boldsymbol{S}_3) = (\boldsymbol{s}_1, \boldsymbol{s}_2, \boldsymbol{s}_3)$ on $\boldsymbol{X}_3 = \boldsymbol{x}_3$ is computed as follows:

$$\left(\boldsymbol{s}_3^T - \boldsymbol{\mu}_3^T\right)\mathbf{B}(\boldsymbol{x}_3 - \boldsymbol{\nu}_3), \tag{7.35}$$

where

$$\boldsymbol{\mu}_t = E(\boldsymbol{S}_t), \boldsymbol{\nu}_t = E(\boldsymbol{X}_t).$$

The total effects of $(\boldsymbol{S}_2, \boldsymbol{S}_3) = (\boldsymbol{s}_2, \boldsymbol{s}_3)$ on $\boldsymbol{X}_3 = \boldsymbol{x}_3$ at $\boldsymbol{S}_1 = \boldsymbol{s}_1$ are calculated by

$$\left(\boldsymbol{s}_3^T - \boldsymbol{\mu}_3^T(\boldsymbol{s}_1)\right)\mathbf{B}(\boldsymbol{x}_3 - \boldsymbol{\nu}_3(\boldsymbol{s}_1)), \tag{7.36}$$

where

$$\boldsymbol{\mu}_3(\boldsymbol{s}_1) = E(\boldsymbol{S}_3|\boldsymbol{S}_1 = \boldsymbol{s}_1), \boldsymbol{\nu}_3(\boldsymbol{s}_1) = E(\boldsymbol{X}_3|\boldsymbol{S}_1 = \boldsymbol{s}_1).$$

The sequence (7.34) is a Markov chain, so the direct effects of $\boldsymbol{S}_t = \boldsymbol{s}_t, t = 1,2$ on $\boldsymbol{X}_3 = \boldsymbol{x}_3$ are zeroes. Subtracting (7.36) from (7.35), it follows that

The total (indirect) effect of $\boldsymbol{S}_1 = \boldsymbol{s}_1$ on $\boldsymbol{X}_3 = \boldsymbol{x}_3$ through $(\boldsymbol{S}_2, \boldsymbol{S}_3) = (\boldsymbol{s}_2, \boldsymbol{s}_3)$

$= $ (the total effect of $(\boldsymbol{S}_1, \boldsymbol{S}_2, \boldsymbol{S}_3) = (\boldsymbol{s}_1, \boldsymbol{s}_2, \boldsymbol{s}_3)$ on $\boldsymbol{X}_3 = \boldsymbol{x}_3$)

$-$ (the total effect of $(\boldsymbol{S}_2, \boldsymbol{S}_3) = (\boldsymbol{s}_2, \boldsymbol{s}_3)$ on $\boldsymbol{X}_3 = \boldsymbol{x}_3$ at $\boldsymbol{S}_1 = \boldsymbol{s}_1$)

$$= \left(\boldsymbol{s}_3^T - \boldsymbol{\mu}_3^T\right)\mathrm{B}(\boldsymbol{x}_3 - \boldsymbol{\nu}_3) - \left(\boldsymbol{s}_3^T - \boldsymbol{\mu}_3^T(\boldsymbol{s}_1)\right)\mathrm{B}(\boldsymbol{x}_3 - \boldsymbol{\nu}_3(\boldsymbol{s}_1)). \tag{7.37}$$

Remark 7.5 Since sequence in (7.34) is a Markov chain, it follows that

The total (indirect) effect of $\boldsymbol{S}_1 = \boldsymbol{s}_1$ on $\boldsymbol{X}_3 = \boldsymbol{x}_3$ through $(\boldsymbol{S}_2, \boldsymbol{S}_3) = (\boldsymbol{s}_2, \boldsymbol{s}_3)$

$=$ The total (indirect) effect of $\boldsymbol{S}_1 = \boldsymbol{s}_1$ on $\boldsymbol{X}_3 = \boldsymbol{x}_3$ through $\boldsymbol{S}_3 = \boldsymbol{s}_3$.

■

Since the total effect of $\boldsymbol{S}_3 = \boldsymbol{s}_3$ on $\boldsymbol{X}_3 = \boldsymbol{x}_3$ at $(\boldsymbol{S}_1, \boldsymbol{S}_2) = (\boldsymbol{s}_1, \boldsymbol{s}_2)$ is given by

$$\left(\boldsymbol{s}_3^T - \boldsymbol{\mu}_3^T(\boldsymbol{s}_2)\right)\mathbf{B}(\boldsymbol{x}_3 - \boldsymbol{\nu}_3(\boldsymbol{s}_2)), \tag{7.38}$$

the total (indirect) effect of $\boldsymbol{S}_2 = \boldsymbol{s}_2$ on $\boldsymbol{X}_3 = \boldsymbol{x}_3$ at $\boldsymbol{S}_1 = \boldsymbol{s}_1$ through $\boldsymbol{S}_3 = \boldsymbol{s}_3$ is calculated by

$$
\begin{aligned}
&(\text{the total effect of } (\boldsymbol{S}_2, \boldsymbol{S}_3) = (\boldsymbol{s}_2, \boldsymbol{s}_3) \text{ on } \boldsymbol{X}_3 = \boldsymbol{x}_3 \text{ at } \boldsymbol{S}_1 = \boldsymbol{s}_1) \\
&\quad - (\text{the total effect of } \boldsymbol{S}_3 = \boldsymbol{s}_3 \text{ on } \boldsymbol{X}_3 = \boldsymbol{x}_3 \text{ at } (\boldsymbol{S}_1, \boldsymbol{S}_2) = (\boldsymbol{s}_1, \boldsymbol{s}_2)) \\
&= \left(\boldsymbol{s}_3^T - \boldsymbol{\mu}_3^T(\boldsymbol{s}_1)\right)\mathbf{B}(\boldsymbol{x}_3 - \boldsymbol{\nu}_3(\boldsymbol{s}_1)) - \left(\boldsymbol{s}_3^T - \boldsymbol{\mu}_3^T(\boldsymbol{s}_2)\right)\mathbf{B}(\boldsymbol{x}_3 - \boldsymbol{\nu}_3(\boldsymbol{s}_2)).
\end{aligned}
$$

Similarly, in the following sequence:

$$S_1 \to S_2 \to S_3,$$

the effects of S_1 and S_2 on S_3 are computed as follows. Since the above sequence is a Markov chain, the direct effects of S_1 on S_3 are zeroes. The total effect of $(\boldsymbol{S}_1, \boldsymbol{S}_2) = (\boldsymbol{s}_1, \boldsymbol{s}_2)$ on $\boldsymbol{S}_3 = \boldsymbol{s}_3$ is given by

$$\left(\boldsymbol{s}_2^T - \boldsymbol{\mu}_2^T\right)\boldsymbol{\Delta}(\boldsymbol{s}_3 - \boldsymbol{\mu}_3). \tag{7.39}$$

The total (direct) effect of $\boldsymbol{S}_2 = \boldsymbol{s}_2$ on $\boldsymbol{S}_3 = \boldsymbol{s}_3$ at $\boldsymbol{S}_1 = \boldsymbol{s}_1$ is calculated by

$$\left(\boldsymbol{s}_2^T - \boldsymbol{\mu}_2^T(\boldsymbol{s}_1)\right)\boldsymbol{\Delta}(\boldsymbol{s}_3 - \boldsymbol{\mu}_3(\boldsymbol{s}_1)). \tag{7.40}$$

From this, the total (indirect) effect of $\boldsymbol{S}_1 = \boldsymbol{s}_1$ on $\boldsymbol{S}_3 = \boldsymbol{s}_3$ through $\boldsymbol{S}_2 = \boldsymbol{s}_2$ is obtained by subtracting (7.40) from (7.39), that is,

$$\left(\boldsymbol{s}_2^T - \boldsymbol{\mu}_2^T\right)\boldsymbol{\Delta}(\boldsymbol{s}_3 - \boldsymbol{\mu}_3) - \left(\boldsymbol{s}_2^T - \boldsymbol{\mu}_2^T(\boldsymbol{s}_1)\right)\boldsymbol{\Delta}(\boldsymbol{s}_3 - \boldsymbol{\mu}_3(\boldsymbol{s}_1)).$$

Finally, we have the total (direct) effect of $\boldsymbol{S}_1 = \boldsymbol{s}_1$ on $\boldsymbol{S}_2 = \boldsymbol{s}_2$ as

$$\left(\boldsymbol{s}_1^T - \boldsymbol{\mu}_1^T\right)\boldsymbol{\Delta}(\boldsymbol{s}_2 - \boldsymbol{\mu}_2).$$

In the next section, the above path analysis for the latent Markov chain is demonstrated by using artificial data.

Remark 7.6 In order to determine regression parameters β_{aj} and δ_{ab} in (7.32) and (7.33), respectively, we have to put a constraint on the parameters. In this section, we set

$$\beta_{j1} = \beta_{1j} = 0,\ j = 1,2,\ldots,J;\ \delta_{a1} = \delta_{1a} = 0,\ a = 1,2,\ldots,A.$$

Then, we have

$$\beta_{aj} = \log\frac{p_{aj}p_{11}}{p_{a1}p_{1j}},\ j = 2,3,\ldots,J;\ \delta_{ab} = \log\frac{m_{ab}m_{11}}{m_{a1}m_{1b}},\ b = 2,3,\ldots,A. \tag{7.41}$$

■

Remark 7.7 In sequence (7.34), we have

The total effect of $\boldsymbol{S}_1 = \boldsymbol{s}_1$ on $\boldsymbol{X}_t = \boldsymbol{x}_t$ through $(\boldsymbol{S}_2, \ldots, \boldsymbol{S}_t) = (\boldsymbol{s}_2, \ldots, \boldsymbol{s}_t)$
$= \left(\boldsymbol{s}_t^T - \boldsymbol{\mu}_t^T\right)\mathbf{B}(\boldsymbol{x}_t - \boldsymbol{v}_t) - \left(\boldsymbol{s}_t^T - \boldsymbol{\mu}_t^T(\boldsymbol{s}_1)\right)\mathbf{B}(\boldsymbol{x}_t - \boldsymbol{v}_t(\boldsymbol{s}_1))$,

and the total effect of $\boldsymbol{S}_u = \boldsymbol{s}_u$ on $\boldsymbol{X}_t = \boldsymbol{x}_t$ at $(\boldsymbol{S}_1, \ldots, \boldsymbol{S}_{u-1}) = (\boldsymbol{s}_1, \ldots, \boldsymbol{s}_{u-1})$ through $\boldsymbol{S}_t = \boldsymbol{s}_t$ is calculated by

$$\left(\boldsymbol{s}_t^T - \boldsymbol{\mu}_t^T(\boldsymbol{s}_{u-1})\right)\mathbf{B}(\boldsymbol{x}_t - \boldsymbol{v}_t(\boldsymbol{s}_{u-1})) - \left(\boldsymbol{s}_t^T - \boldsymbol{\mu}_t^T(\boldsymbol{s}_u)\right)\mathbf{B}(\boldsymbol{x}_t - \boldsymbol{v}_t(\boldsymbol{s}_u)), \qquad (7.42)$$

where

$$\boldsymbol{\mu}_t^T(\boldsymbol{s}_k) = \boldsymbol{s}_k^T\boldsymbol{M}^{t-k},\ \boldsymbol{v}_t(\boldsymbol{s}_k) = \boldsymbol{s}_k^T\boldsymbol{M}^{t-k}\boldsymbol{P}, k = 1,2,\ldots,t-1. \qquad (7.43)$$

Since state vector $\boldsymbol{s}_k^T$ is one of the following A unit vectors:

$$(1,0,\ldots,0), (0,1,0,\ldots,0), (0,\ldots,0.1),$$

the conditional distribution vectors $\boldsymbol{\mu}_t^T(\boldsymbol{s}_k)$ and $\boldsymbol{v}_t(\boldsymbol{s}_k)$ are given by appropriate rows of matrices $\boldsymbol{M}^{t-k}$ and $\boldsymbol{M}^{t-k}\boldsymbol{P}$, respectively. For example, for $\boldsymbol{s}_k^T = (1,0,\ldots,0)$, $\boldsymbol{\mu}_t^T(\boldsymbol{s}_k)$ and $\boldsymbol{v}_t(\boldsymbol{s}_k)$ are obtained as the first rows of the matrices, respectively. If the Markov chain with transition matrix $\boldsymbol{M}$ is irreducible and recurrent, we have

$$\boldsymbol{M}^{t-k} \to \boldsymbol{\Pi} = \begin{pmatrix} \pi_1 & \pi_2 & \cdots & \pi_A \\ \pi_1 & \pi_2 & \cdots & \pi_A \\ \vdots & \vdots & \cdots & \vdots \\ \pi_1 & \pi_2 & \cdots & \pi_A \end{pmatrix}, \text{ as } t \to \infty, \qquad (7.44)$$

where

$$\pi_a \geq 0, a = 1,2,\ldots,A; \sum_{a=1}^{A}\pi_a = 1.$$

Hence, for fixed integer u the effects in (7.42) tend to zeroes as $t \to \infty$.

7.7 Numerical Illustration II

In the model treated in the previous section, for $A = J = 3$, let the response probability and latent transition matrices be set as

$$\begin{pmatrix} p_{11} & p_{12} & p_{13} \\ p_{21} & p_{22} & p_{23} \\ p_{31} & p_{32} & p_{33} \end{pmatrix} = \begin{pmatrix} 0.8 & 0.1 & 0.1 \\ 0.2 & 0.7 & 0.1 \\ 0.1 & 0.2 & 0.7 \end{pmatrix},$$

$$\begin{pmatrix} m_{11} & m_{12} & m_{13} \\ m_{21} & m_{22} & m_{23} \\ m_{31} & m_{32} & m_{33} \end{pmatrix} = \begin{pmatrix} 0.6 & 0.3 & 0.1 \\ 0.2 & 0.7 & 0.1 \\ 0.1 & 0.3 & 0.6 \end{pmatrix}, \tag{7.45}$$

respectively; and for the initial distribution of latent state $\boldsymbol{S}_1 = (S_{11}, S_{12}, S_{13})$,

$$\boldsymbol{\mu}_1 = (\mu_{11}, \mu_{12}, \mu_{13})^T = (0.3,0.6,0.1)^T. \tag{7.46}$$

The path analysis in Sect. 7.6 is demonstrated. According to the eigenvalue decomposition of transition matrix (7.45), we have

$$\begin{pmatrix} m_{11} & m_{12} & m_{13} \\ m_{21} & m_{22} & m_{23} \\ m_{31} & m_{32} & m_{33} \end{pmatrix} = \begin{pmatrix} 0.577 & 0.236 & 0.577 \\ 0.577 & 0.236 & -0.577 \\ 0.577 & -0.943 & 0.577 \end{pmatrix}$$
$$\begin{pmatrix} 1 & 0 & 0 \\ 0 & 0.5 & 0 \\ 0 & 0 & 0.4 \end{pmatrix} \begin{pmatrix} 0.577 & 0.236 & 0.577 \\ 0.577 & 0.236 & -0.577 \\ 0.577 & -0.943 & 0.577 \end{pmatrix}^{-1},$$

and thus, for integer t, it follows that

$$\begin{pmatrix} m_{11} & m_{12} & m_{13} \\ m_{21} & m_{22} & m_{23} \\ m_{31} & m_{32} & m_{33} \end{pmatrix}^t = \begin{pmatrix} 0.577 & 0.236 & 0.577 \\ 0.577 & 0.236 & -0.577 \\ 0.577 & -0.943 & 0.577 \end{pmatrix} \begin{pmatrix} 1 & 0 & 0 \\ 0 & 0.5^t & 0 \\ 0 & 0 & 0.4^t \end{pmatrix} \begin{pmatrix} 0.577 & 0.236 & 0.577 \\ 0.577 & 0.236 & -0.577 \\ 0.577 & -0.943 & 0.577 \end{pmatrix}^{-1},$$
$$= \begin{pmatrix} 0.2 \times 0.5^t + 0.5 \times 0.4^t + 0.3 & -0.5 \times 0.4^t + 0.5 & -0.2 \times 0.5^t + 0.2 \\ 0.2 \times 0.5^t - 0.5 \times 0.4^t + 0.3 & 0.5 \times 0.4^t + 0.5 & -0.2 \times 0.5^t + 0.2 \\ 0.5 \times 0.4^t - 0.8 \times 0.5^t + 0.3 & -0.5 \times 0.4^t + 0.5 & 0.8 \times 0.5^t + 0.2 \end{pmatrix}.$$

The above matrices imply the conditional distribution of $\boldsymbol{S}_{t+u} = (S_{t+u,1}, S_{t+u,2}, S_{t+u,3})^T$ given $\boldsymbol{S}_u$. As shown in (7.44), as integer t goes to infinity, we have

$$\begin{pmatrix} m_{11} & m_{12} & m_{13} \\ m_{21} & m_{22} & m_{23} \\ m_{31} & m_{32} & m_{33} \end{pmatrix}^t \to \begin{pmatrix} 0.3 & 0.5 & 0.2 \\ 0.3 & 0.5 & 0.2 \\ 0.3 & 0.5 & 0.2 \end{pmatrix}.$$

At time t, the distribution of $\boldsymbol{S}_t = (S_{t1}, S_{t2}, S_{t3},)^T$ is calculated as

$$\boldsymbol{\mu}_t^T = (\mu_{t1}, \mu_{t2}, \mu_{t3}) = \boldsymbol{\mu}_1^T \begin{pmatrix} m_{11} & m_{12} & m_{13} \\ m_{21} & m_{22} & m_{23} \\ m_{31} & m_{32} & m_{33} \end{pmatrix}^{t-1}$$
$$= \left(0.1 \times 0.5^{t-1} - 0.1 \times 0.4^{t-1} + 0.3, 0.1 \times 0.4^{t-1} + 0.5, 0.2 - 0.1 \times 0.5^{t-1}\right),$$

and the marginal distribution of $\boldsymbol{X}_t = (X_{t1}, X_{t2}, X_{t3},)$ is calculated as

$$\boldsymbol{\nu}_t^T = \boldsymbol{\mu}_1^T \begin{pmatrix} m_{11} & m_{12} & m_{13} \\ m_{21} & m_{22} & m_{23} \\ m_{31} & m_{32} & m_{33} \end{pmatrix}^{t-1} \begin{pmatrix} p_{11} & p_{12} & p_{13} \\ p_{21} & p_{22} & p_{23} \\ p_{31} & p_{32} & p_{33} \end{pmatrix}$$
$$= \left(0.07 \times 0.5^{t-1} - 0.06 \times 0.4^{t-1} + 0.36, 0.06 \times 0.4^{t-1} - 0.02 \times 0.5^{t-1} + 0.44, 0.2 - 0.05 \times 0.5^{t-1}\right).$$

By using (7.41), regression matrix B can be obtained as follows:

$$\mathrm{B} = \begin{pmatrix} 0 & 0 & 0 \\ 0 & 3.332 & 1.386 \\ 0 & 2.773 & 4.025 \end{pmatrix}.$$

In the present example, state vector $\boldsymbol{s}_k^T$ in (7.43) is one of the following unit vectors:

$$(1,0,0), (0,1,0), (0,0.1).$$

To demonstrate the approach explained in the previous section, for $t = 3$ in (7.34), the effects of latent variables $S_i, i = 1,2,3$ on X_3 are calculated. First, the total effects of the latent variables, $\left(\boldsymbol{s}_3^T - \boldsymbol{\mu}_3^T\right)\mathbf{B}(\boldsymbol{x}_3 - \boldsymbol{\nu}_3)$, in (7.35) are computed by using the following matrix calculation:

$$\begin{pmatrix} 0.7 & -0.6 & -0.1 \\ -0.3 & 0.4 & -0.1 \\ -0.3 & -0.6 & 0.9 \end{pmatrix} \begin{pmatrix} 0 & -2.079 & -2.079 \\ 0 & 1.253 & -0.693 \\ 0 & 0.693 & 1.946 \end{pmatrix} \begin{pmatrix} 0.632 & -0.368 & -0.368 \\ -0.427 & 0.573 & -0.427 \\ -0.205 & -0.205 & 0.795 \end{pmatrix}$$
$$= \begin{pmatrix} 1.225 & -1.051 & -0.009 \\ -0.482 & 0.574 & -0.330 \\ -0.784 & -0.288 & 2.007 \end{pmatrix}. \tag{7.47}$$

The above effects are those of $(\boldsymbol{S}_1, \boldsymbol{S}_2, \boldsymbol{S}_3)$ on $\boldsymbol{X}_3$, for example, for $\boldsymbol{S}_3 = (0,1,0)^T$, the effect on $\boldsymbol{X}_3 = (0,0,1)^T$ is -0.330, which is in the second row and the third column. Since sequence S_1, S_2, S_3, and X_3 is a Markov chain, the effects in (7.47) are independent of latent states S_1 and S_2. The mean total effect of elements in (7.47) can be obtained as 0.633. Hence, the entropy coefficient of determination of explanatory variables $\boldsymbol{S}_i, i = 1,2,3$ for response variable $\boldsymbol{X}_3$ is

$$ECD((\boldsymbol{S}_1, \boldsymbol{S}_2, \boldsymbol{S}_3), \boldsymbol{X}_3) = ECD((\boldsymbol{S}_3, \boldsymbol{X}_3), \boldsymbol{X}_3) = \frac{0.633}{0.633 + 1} = 0.387.$$

From path analysis [9], the above quantity is the standardized summary total effect of $(\boldsymbol{S}_1, \boldsymbol{S}_2, \boldsymbol{S}_3)$ on $\boldsymbol{X}_3$, and is denoted by $e_T((\boldsymbol{S}_1, \boldsymbol{S}_2, \boldsymbol{S}_3) \to \boldsymbol{X}_3)$. Similarly, from (7.36) the total effects of $(\boldsymbol{S}_2, \boldsymbol{S}_3) = (\boldsymbol{s}_2, \boldsymbol{s}_3)$ on $\boldsymbol{X}_3 = \boldsymbol{x}_3$ are given as follows:

$$\begin{pmatrix} 0.892 & -0.922 & -0294 \\ -0.595 & 0.922 & -0.394 \\ -0.891 & 0.066 & 1.949 \end{pmatrix} \text{ at } \boldsymbol{S}_1 = (1,0,0)^T; \tag{7.48a}$$

$$\begin{pmatrix} 1.355 & -0.994 & -0.053 \\ -0.451 & 0.532 & -0.473 \\ -0.694 & 0.270 & 1.924 \end{pmatrix} \text{ at } \boldsymbol{S}_1 = (0,1,0)^T; \tag{7.48b}$$

$$\begin{pmatrix} 1.729 & -0.7804 & -0.464 \\ -0.045 & 0.775 & -0.855 \\ -0.727 & 0.463 & 1.106 \end{pmatrix} \text{ at } \boldsymbol{S}_1 = (0,0,1)^T. \tag{7.48c}$$

In the above effects, for example, for $\boldsymbol{s}_1 = (0,1,0)^T$, $\boldsymbol{s}_3 = (0,0,1)^T$, and $\boldsymbol{x}_3 = (1,0,0)^T$, the total effect is -0.694 which is in the third row and first column of the matrix (7.48b). Sequence S_1, S_2, S_3, and X_3 is a Markov chain, so the above effects of $(\boldsymbol{S}_2, \boldsymbol{S}_3) = (\boldsymbol{s}_2, \boldsymbol{s}_3)$ on $\boldsymbol{X}_3 = \boldsymbol{x}_3$ are independent of latent state S_2. The conditional means of the above effects given $\boldsymbol{S}_1 = \boldsymbol{s}_1$ are obtained by

$$E\left(\left(\boldsymbol{S}_3^T - \boldsymbol{\mu}_3^T(\boldsymbol{s}_1)\right)\mathbf{B}(\boldsymbol{X}_3 - \boldsymbol{\nu}_3(\boldsymbol{s}_1))|\boldsymbol{s}_1\right) = \begin{cases} 0.652 \left(\boldsymbol{S}_1 = (1,0,0)^T\right) \\ 0.584 \left(\boldsymbol{S}_1 = (0,1,0)^T\right) \\ 0.658 \left(\boldsymbol{S}_1 = (0,0,1)^T\right) \end{cases}.$$

Since the distribution of $\boldsymbol{S}_1$ is in (7.46), we have that the mean total effect of $(\boldsymbol{S}_2, \boldsymbol{S}_3)$ on $\boldsymbol{X}_3$ is computes by

$$0.652 \times 0.3 + 0.584 \times 0.6 + 0.658 \times 0.3 = 0.612. \tag{7.49}$$

Subtracting the effects in (7.48a)–(7.48c) from those in (7.47), from (7.37), the total (indirect) effects of $\boldsymbol{S}_1 = \boldsymbol{s}_1$ on $\boldsymbol{X}_3 = \boldsymbol{x}_3$ through $(\boldsymbol{S}_2, \boldsymbol{S}_3) = (\boldsymbol{s}_2, \boldsymbol{s}_3)$ are obtained by

$$\begin{pmatrix} 0.334 & -0.127 & 0.285 \\ 0.113 & -0.348 & 0.064 \\ 0.107 & -0.354 & 0.058 \end{pmatrix} \left(\boldsymbol{S}_1 = (1,0,0)^T\right); \tag{7.50a}$$

$$\begin{pmatrix} -0.130 & -0.058 & 0.044 \\ -0.031 & 0.041 & 0.143 \\ -0.090 & -0.018 & 0.083 \end{pmatrix} \left(\boldsymbol{S}_1 = (0,1,0)^T\right); \tag{7.50b}$$

$$\begin{pmatrix} -0.503 & -0.271 & 0.455 \\ -0.433 & -0.201 & 0.525 \\ -0.057 & 0.175 & 0.901 \end{pmatrix} \left(\boldsymbol{S}_1 = (0,0,1)^T\right). \tag{7.50c}$$

In the above matrices, for $\boldsymbol{s}_1 = (0,0,1)^T$, $\boldsymbol{s}_3 = (0,0,1)^T$, and $\boldsymbol{x}_3 = (0,1,0)^T$, the total effect is calculated in the third row the second column in matrix (7.50c) and given by 0.175. The above effects are independent of latent states S_2, because sequence S_1, S_2, S_3, and X_3 is a Markov chain. The summary total effect of $\boldsymbol{S}_1$ on $\boldsymbol{X}_3$ through $(\boldsymbol{S}_2, \boldsymbol{S}_3)$ is given by subtracting that of $(\boldsymbol{S}_2, \boldsymbol{S}_3)$, i.e., 0.612, from that of $(\boldsymbol{S}_1, \boldsymbol{S}_2, \boldsymbol{S}_3)$, i.e., 0.633. Hence, the effect is

$$0.633 - 0.612 = 0.021.$$

From this, the standardized summary total (indirect) effect of $\boldsymbol{S}_1$ on $\boldsymbol{X}_3$ is computed as

$$e_T(\boldsymbol{S}_1 \to \boldsymbol{X}_3) = \frac{0.021}{0.633 + 1} = 0.013.$$

By using (7.38), the total (direct) effect of $\boldsymbol{S}_3 = \boldsymbol{s}_3$ on $\boldsymbol{X}_3 = \boldsymbol{x}_3$ at $(\boldsymbol{S}_1, \boldsymbol{S}_2) = (\boldsymbol{s}_1, \boldsymbol{s}_2)$ is calculated as follows:

$$\begin{pmatrix} 0.501 & -0.776 & -0.317 \\ -0.687 & 1.368 & -0.119 \\ -0.947 & 0.549 & 2.260 \end{pmatrix} \text{ at } \boldsymbol{S}_2 = (1,0,0)^T, \tag{7.51a}$$

$$\begin{pmatrix} 1.603 & -1.007 & 0.230 \\ -0.385 & 0.337 & -0.372 \\ -0.511 & -0.348 & 2.142 \end{pmatrix} \text{ at } \boldsymbol{S}_2 = (0,1,0)^T, \tag{7.51b}$$

$$\begin{pmatrix} 2.208 & -0.455 & -0.623 \\ 0.437 & 1.106 & -1.008 \\ -0.587 & -0.477 & 0.608 \end{pmatrix} \text{ at } \boldsymbol{S}_2 = (0,0,1)^T. \tag{7.51c}$$

In the above effects, for $\boldsymbol{s}_2 = (0,1,0)^T$, $\boldsymbol{s}_3 = (1,0,0)^T$, and $\boldsymbol{x}_3 = (0,0,1)^T$, the effect is in the first row and the third column in matrix (7.51b) and is 0.230. Since the marginal distribution of $\boldsymbol{S}_2$ is given by

$$\boldsymbol{\mu}_2^T = \boldsymbol{\mu}_1^T \boldsymbol{M} = (0.31,0.54,0.15),$$

from the same way as in (7.49), we have the conditional mean of the effects in (7.51a)–(7.51c) as

$$0.577 \times 0.31 + 0.464 \times 0.54 + 0.557 \times 0.15 = 0.513.$$

From this we have

$$e_T(\boldsymbol{S}_3 \to \boldsymbol{X}_3) = \frac{0.513}{0.633 + 1} = 0.314.$$

The total (indirect) effects of $\boldsymbol{S}_2 = \boldsymbol{s}_2$ on $\boldsymbol{X}_3 = \boldsymbol{x}_3$ at $\boldsymbol{S}_1 = \boldsymbol{s}_1$ through $\boldsymbol{S}_3 = \boldsymbol{s}_3$ are calculated as follows. Subtracting the effects in (7.51a) from those in (7.48a), we have the total (indirect) effects of $\boldsymbol{S}_2 = \boldsymbol{s}_2$ on $\boldsymbol{X}_3 = \boldsymbol{x}_3$ at $\boldsymbol{S}_1 = (1,0,0)^T$ through $\boldsymbol{S}_3 = (1,0,0)^T$:

$$\begin{pmatrix} 0.390 & -0.148 & 0.023 \\ 0.092 & -0.446 & -0.275 \\ 0.056 & -0.482 & -0.312 \end{pmatrix} \text{ at } \boldsymbol{S}_1 = (1,0,0)^T \text{ through } \boldsymbol{S}_3 = (1,0,0)^T.$$

In the above effects, the effect of $\boldsymbol{S}_2 = (0,1,0)^T$ on $\boldsymbol{X}_3 = (1,0,0)$ is in the second row and the first column, and is given by 0.092. Similarly, the other effects can be calculated. For example, the total (indirect) effects of $\boldsymbol{S}_2 = \boldsymbol{s}_2$ on $\boldsymbol{X}_3 = \boldsymbol{x}_3$ at $\boldsymbol{S}_1 = (1,0,0)^T$ through $\boldsymbol{S}_3 = (0,1,0)^T$ are obtained by subtracting the effects in (7.51a) from those in (7.48b), and we have

$$\begin{pmatrix} 0.854 & -0.218 & 0.264 \\ 0.236 & -0.836 & -0.354 \\ 0.253 & -0.279 & -0.336 \end{pmatrix} \text{ at } \boldsymbol{S}_1 = (1,0,0)^T \text{ through } \boldsymbol{S}_3 = (0,1,0)^T.$$

From the effects calculated above, the summary effect is calculated as

$$0.612 - 0.513 = 0.099.$$

and the standardized summary total (indirect) effect is given by

$$e_T(\boldsymbol{S}_2 \to \boldsymbol{X}_3) = \frac{0.099}{0.633 + 1} = 0.061.$$

As shown above, calculation of the effects of latent variables $\boldsymbol{S}_t, t = 1,2,3$ on $\boldsymbol{X}_3$ has been demonstrated. Similarly, the effects of $\boldsymbol{S}_i$ on $\boldsymbol{S}_j$ can also be computed. The summary total effects $e_T(\boldsymbol{S}_t \to \boldsymbol{X}_3)$ in this example, as expected from the following relation:

$$S_1 \to S_2 \to S_3 \to X_3;$$

the following inequality holds true:

$$e_T(S_1 \to X_3) < e_T(S_2 \to X_3) < e_T(S_3 \to X_3).$$

7.8 Discussion

In this chapter, path analysis has been made in latent class models, i.e., multiple-indicator, multiple-cause models and the latent Markov chain model. In path analysis, it is critical how the effects of variables are measured and also how the total effects of variables are decomposed into the sums of the direct and indirect effects. Although the approach is significant for discussing causal systems of categorical variables, path analysis of categorical variables is more complicated than that of continuous variables, because in the former analysis the effects of categories of parent variables on those of descendant variables have to be calculated. In order to assess the effects and to summarize them, in this chapter, an entropy-based path analysis [9] has been applied to latent class models in a GLM framework. In the approach, the total and direct effects are defined through log odds ratio and the effects can be interpreted in information (entropy). From this, although the indirect effects are defined by subtracting the direct effects from the total effects, the effects can also be interpreted in information. This point is significant for putting the analysis into practice. Measuring pathway effects based on the method of path analysis is important as well, and further development of pathway effect analysis is left to readers. Moreover, applications of the present approach to practical latent class analyses are also expected in future studies.

References

1. Albert, J. M., & Nelson, S. (2011). Generalized causal mediation analysis. *Biometrics,* 1028–1038.
2. Bentler, P. M., & Weeks, D. B. (1980). Linear structural equations with latent variables. *Psychometrika, 45,* 289–308.
3. Christoferson, A. (1975). Factor analysis of dichotomous variables. *Psychometrika, 40,* 5–31.
4. Eshima, N., & Tabata, M. (1999). Effect analysis in loglinear model approach to path analysis of categorical variables. *Behaviormetrika, 26,* 221–233.
5. Eshima, N., & Tabata, M. (2007). Entropy correlation coefficient for measuring predictive power of generalized linear models. *Statistics and Probability Letters, 77,* 588–593.
6. Eshima, N., & Tabata, M. (2010). Entropy coefficient of determination for generalized linear models. *Computational Statistics and Data Analysis, 54,* 1381–1389.
7. Eshima, N., Asano, C., & Obana, E. (1990). A latent class model for assessing learning structures. *Behaviormetrika, 28,* 23–35.
8. Eshima, N., Tabata, M., & Geng, Z. (2001). Path analysis with logistic regression models: Effect analysis of fully recursive causal systems of categorical variables. *Journal of the Japan Statistical Society, 31,* 1–14.

9. Eshima, N., Tabata, M., Borroni, C. G., & Kano, Y. (2015). An entropy-based approach to path analysis of structural generalized linear models: A basic idea. *Entropy, 17*, 5117–5132.
10. Fienberg, S. E. (1991). *The analysis of cross-classified categorical data* (2nd ed.). Cambridge, England: The MIT Press.
11. Goodman, L. A. (1973b). The analysis of multidimensional contingency tables when some variables are posterior to others: A modified path analysis approach. *Biometrika, 60*, 179–192.
12. Goodman, L. A. (1973a). Causal analysis of data from panel studies and other kinds of surveys. *American Journal of Sociology, 78*, 1135–1191.
13. Goodman, L. A. (1974). The analysis of systems of qualitative variables when some of the variables are unidentifiable: Part I. A modified latent structure approach. *American Journal of Sociology, 79*, 1179–1259.
14. Hagenaars, J. A. (1998). Categorical causal modeling: Latent class analysis and directed loglinear models with latent variables. *Sociological Methods & Research, 26*, 436–489.
15. Jöreskog, K.G., & Sörbom, D. (1996). *LISREL8: user's reference guide* (2nd ed.). Chicago: Scientific Software International.
16. Kuha, J., & Goldthorpe, J. H. (2010). Path analysis for discrete variables: The role of education in social mobility. *Journal of Royal Statistical Society, A, 173*, 351–369.
17. Lazarsfeld, P. F. (1948). The use of panels in social research. *Proceedings of the American Philosophical Society, 92*, 405–410.
18. Macready, G. B. (1982). The use of latent class models for assessing prerequisite relations and transference among traits, Psychometrika, 47, 477-488.
19. McCullagh, P., & Nelder, J. A. (1989). *Generalized linear models* (2nd ed.). London: Chapman and Hall.
20. Muthen, B. (1978). Contribution of factor analysis of dichotomous variables. *Psychometrika, 43*, 551–560.
21. Muthen, B. (1984). A general structural equation model with dichotomous ordered categorical and continuous latent variable indicators. *Psychometrika, 49*, 114–132.
22. Nelder, J. A., & Wedderburn, R. W. M. (1972). Generalized linear model. *Journal of the Royal Statistical Society A, 135*, 370–384.
23. Owston, R. D. (1979). A maximum likelihood approach to the "test of inclusion." *Psychometrika, 44*, 421–425.
24. White, R. T., & Clark, R. M. (1973). A test of inclusion which allows for errors of measurement. *Psychometrika, 38*, 77–86.
25. Wright, S. (1934). The method of path coefficients. *The Annals of Mathematical Statistics, 5*, 161–215.

www.ingramcontent.com/pod-product-compliance
Ingram Content Group UK Ltd.
Pitfield, Milton Keynes, MK11 3LW, UK
UKHW020817290926
15470UKWH00003B/1220
* 9 7 8 9 8 1 1 9 0 9 7 1 9 *